THAT RIBBON OF HIGHWAY II:

Highway 99 From the State Capital to the Mexican Border

By Jill Livingston

Photographs By Kathryn Golden Maloof

Cover photograph is of Golden State Highway, a piece of former US Highway 99 below Pyramid Lake in Los Angeles County.

All photos without listed credits are contemporary photos by Kathryn Golden Maloof.
Drawings by Kathryn Golden Maloof.
Maps by Jill Livingston.
Book design by Living Gold Press.

Book title is from the song *This Land is Your Land* by Woodrow Wilson Guthrie.

LIVING GOLD PRESS

P.O. Box 2
Klamath River
CA 96050
www.livinggoldpress.com

Comments and corrections encouraged.

The road topped the crest of the Siskiyous, cut through a 4000 foot mountain pass chiseled out of solid rock, headed sharply downhill and south, and in a couple of miles crossed into a part of California sitting on the southern threshold of the conifer-covered Pacific Northwest. Over 900 miles later it ended in a sweltering desert valley on the border of another country. Inbetween, it bisected the state neatly into East and West. On its way south, it crossed mountain ranges, wound through tortuous river canyons, bee-lined across broad valleys dotted with magnificent oaks, passed through fertile fields and orchards, made its mark on innumerable settlements, large and small... This was Highway 99.

from *That Ribbon of Highway: Highway 99 From the Oregon Border to the State Capital*

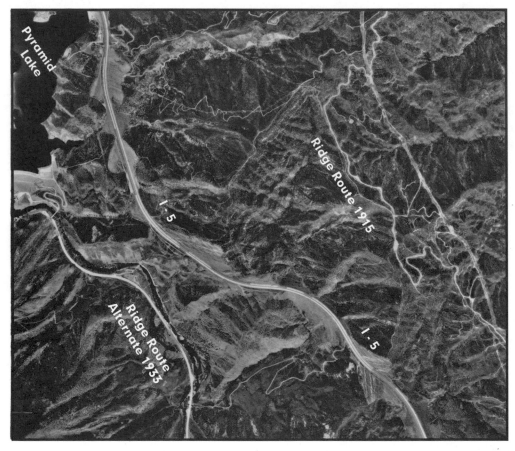

Pyramid Lake

Ridge Route 1915

I-5

Ridge Route Alternate 1933

I-5

Three successive versions of US Highway 99 crossing the Tehachapi Mountains can be seen on this USGS aerial view from 1994. The narrow and windy 1915 road on the right was the first one dubbed The Ridge Route.

Its 1933 replacement, the Ridge Route Alternate, is on the left. After WWII this road was widened into a four lane expressway. The section through the scenic Piru Gorge was submerged under Pyramid Lake in 1970.

Modern I-5 is in the middle. The "99" shields are long gone, but this road is still referred to as "the Ridge Route."

CONTENTS

(CalTrans)

CHAPTER 1

A CONTINUOUS SLAB OF CONCRETE

My bare legs glued themselves to the smooth naugahyde as the long, sweltering, endless green valley whizzed by the back seat window. An ice cold soda enjoyed in a shady roadside fruit stand provided a temporary modicum of relief. Ripe juicy peaches to go, yum. Or, purchase an orange. Do not peel. Squeeze it, massage it, roll it around until it feels soft and pulpy. Poke a hole in the top with your finger, insert a straw if you have one, and voila! Instant orange juice in its own container. At least a full fifteen minutes' backseat entertainment.

Indescribably delicious to be sure, but the dripping juice added another layer of stickiness. However, a child's tolerance of stickiness is generally greater than that of an adult. And so I stoically endured the long journey down Highway 99, one of many. Relief would come at the end of the day in the cooling waters of a Bakersfield motel swimming pool...

Then and Now

Most of the fruit stands are gone. The mom-and-pop motels with their brilliant neon signs and fat flashing arrows are long past their prime. Service stations no longer dispense free maps or free service. The child is well into middle age. The millennium approaches.

Yet something more tangible than the indelible memories of youth does indeed remain of Highway 99. As a surviving relic of recent history, there are still sights to see along the old road.

Long stretches of this venerable highway are intact. In its north-south transect through the center of the state, the heartland of California, Highway 99 seems to be an apt representative of the social history that shaped us in the first three quarters of the 20th century.

Highway 99 attests to the time in our past when getting from one place to the next was a journey to be enjoyed. Not the stress-filled race from here to there, punctuated by occasional stops at homogeneous off-ramp conglomerations of gasoline, fast food, and motel chains, that is the typical mode of travel today.

As Americans, so wrapped up are we in our cars, our roads, the businesses that serve and entertain us along our roads, so obsessed are we with just going someplace in general, that the evolution of highways can be considered a major component of our social evolution. And nowhere more so than in California, where

Initially, Highway 99 was paved in a single slab of concrete only 15' wide, such as this stretch in Merced County. By the late 1920s all but 100 miles of the Canada-to-Mexico route had been paved and the road was declared "the longest continuously improved highway in the country." (CalTrans)

distances from city to city are far, where many of the cities themselves took shape around the automobile culture.

The Thrill of Discovery

There seems to be a revival of interest in history as it relates to automobiles, motorists, highways, and roadside culture and architecture. Where lies the appeal? What is the value?

Obviously there is no one answer to this question. Some of the fun lies in putting together the puzzle pieces that are left of the now incomplete, renamed or unnamed Highway 99.

Where the interstate cuts a straight and smooth swath across the landscape, bypassing city and town, the old highway looped and bumped its way up and down the western edge of the continent, connecting previously isolated cities and towns and rural areas. By rejoining the severed fragments of the once continuous ribbon,

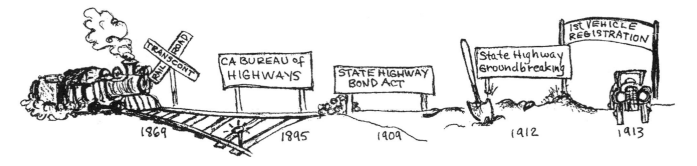

4

by searching for roadside relics, we seek physical pieces of evidence that help bring to light a significant facet of our recent, shared experiences. Modest though the artifacts may be, they instill a sense of history, a feel for the past. We are roadside archeologists caught up in the thrill of discovery.

Highway History in a Nutshell

In a nutshell (and make that a California walnut, almond, or pistachio), the automobile era in California began in 1895 with the formation of the California Burcau of Highways. (This subject is covered in more detail in *That Ribbon of Highway: Highway 99 from the Oregon Border to the State Capital*, the first book in this series.) This new bureaucracy was charged with planning a state-wide road system. The planning was done and the future system mapped out, all at a time when horses and wagons were the dominant form of transportation. But there was no money to make anything of all the great plans.

Meanwhile, a grass roots movement took hold called the Good Roads Movement. Bicyclists (for bicycling was extremely popular around the turn of the century) were the first to demand better roads, to send lobbyists to Sacramento and Washington. Driving an automobile was still a rich man's hobby. That soon changed.

In a remarkably short period of time after the Model T was introduced by the Ford Motor Company in 1909, car ownership became democratized, just as Henry Ford had set out to do. Cars were suddenly affordable. There was no stopping the hoards of motorists that then took to the roads, and they quickly grew tired of bogging down in the mud.

In 1909, the first State Highway Bond Act was easily voted in, establishing the State Highway System. The first State Highway groundbreaking took place in 1912. When the money ran out, other Bond Acts followed in 1916 and 1919. A two cent fuel tax was initiated in 1923 to fund road construction and maintenance.

The Federal government started contributing money in 1916. The roads eligible for federal funds were limited to not more than 3% of total highway miles in each

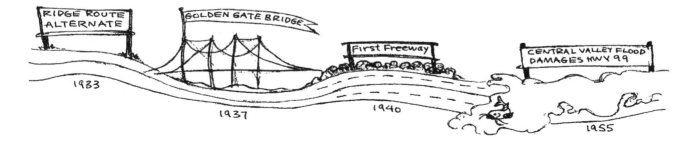

RIDGE ROUTE ALTERNATE
1933

GOLDEN GATE BRIDGE
1937

First Freeway
1940

CENTRAL VALLEY FLOOD DAMAGES HWY 99
1955

state and they were still considered State Highways. They did not officially become US Highways, such as US Highway 99, until 1926.

At that time, the current highway numbering system was adopted. North-south routes were given odd numbers, east-west were given even numbers. Highway numbers, with a few exceptions, increased from the east coast to the west coast.

The newly chosen US Highways, which in essence were simply the most important State Highways, continued to be identified by their various State Highway numbers as well, especially by the Dept. of Highways. Motorists persisted in calling the highways by their more interesting localized names. Highway 99 was (still is) the Ridge Route or the Grapevine over the mountains north of Los Angeles, the Valley Route or Golden State Highway through the Central Valley, the Pacific Highway farther north. But after the US Highway shields went up starting in 1928, "99" achieved an identity of its own.

With the influx of bond and government money from 1909 onward, existing roads that started out as wagon roads were improved, gaps connected by new

construction, bridges built, and everything in due time was paved. The future Highway 99, along with coastal 101, was one of the two main north-south lines. The road building tools used were primitive: men with wheel barrows, mules pulling Fresno scrapers, steam shovels only in extreme conditions due to their expense and rarity.

The number of motorists multiplied as rapidly as the cars rolled off of the newly created assembly lines and installment buying was introduced. That initial 15 foot wide strip of concrete was poured in a single slab. As needed, the highway was widened, another lane or two added. Highways of the twenties and thirties were double or triple-slab. Eventually the concrete was paved over with asphalt. A 1928 U.S. Dept. of Agriculture press release touted Highway 99 as "the longest continuously improved highway in the country."

Man power and beasts of burden, such as these hard at work on the 1915 Ridge Route, get the credit for the first rash of highway construction in the teens. The Highway Department owned their own stock. (CalTrans)

In the postwar babyboom years the increase in motorists was even more astronomical. The Interstate Highway System, a nationwide web of improved highways built to high standards, was a concept whose appeal grew as the traffic worsened and the funding trickled in. It was not fully funded until 1956. Curiously, one mile out of five was required to be straight enough on which to land an airplane, a Cold War-era defense measure. President Eisenhower had been impressed by the efficient autobahns he saw in WWII Germany and was a key player in its implementation.

Thus, the two or three-lane highway evolved into an "expressway," four lanes and sometimes divided but still allowing for cross traffic. The expressway evolved into a "freeway," with all access controlled by on and offramps. And the two-lane highway, so grand at its birth, lost its identity.

What remains of the original Highway 99 is now

San Joaquin County, 1931. Heavy equipment was commonly used in highway construction starting in the 1920s. Some of it was WWI surplus.(CalTrans)

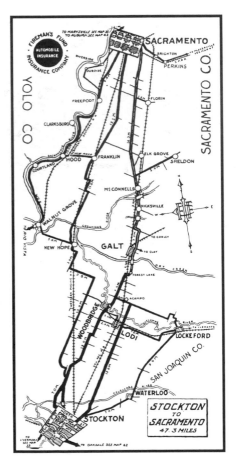

1921 map from Automobile Tour Book of California, *Fireman's Fund Insurance of California.*

either incorporated into the freeway (buried underneath it, or "subsumed" in highwayspeak), relegated to frontage road or business loop, or simply abandoned.

In these disguises, seeking what highway and roadside relics remain can provide a rewarding glimpse back through a window into the past. The recent past. A past experienced not by ancient, faceless ancestors, but by ourselves or our parents or grandparents.

Further on Down the Road...

The flickering windshield images have stayed with me for forty plus years. The stretch of Highway 99 from near Sacramento to southern California was in fact the portion of the road in which the oldest memories lie.

The first leg of the journey down California, from the Oregon border to Sacramento (as covered in the first book in this series) deposited us right at the State Capitol building. The highway led us into, and will shortly take us out of, the center of the capital

10

city. Despite our long journey over mountain, through canyon, across valley, in Sacramento we are still only one third of the way down the state and to the end (or beginning) of the highway in Calexico on the Mexican border.

We have yet to travel down a very long, straight valley, across a range of mountains, wend through a metropolis, over a windy pass, then decend into and across a vast desert, finally stopping at the

This vintage sign in Merced still bears a familiar but outdated slogan.

chain link fence that separates us from our neighbors.

All of these places are a part of this thing called California, and all are populated by more than a handful of people by the grace of rearranged water. All are united primarily by the strip of pavement that grew more crowded and wider over the years, that brought workers to the fields, crops to the market, motorists to the motels and eateries and roadside attractions, settlers to the towns. That in turn shaped the landscape itself by inspiring sprawling cities, drive-in restaurants and theaters, motel rows, malls.

But let us not go that far toward the present. Piecing together remnants of the old highway is a way to piece together the past. We return to a time when growth was considered a good thing, when families took leisurely road trips, when Dad could easily work on the car with just a few tools. There is from where we came.

Along Highway 99, Merced County in 1937. In some places the roadside seems to have changed very little, although the US Highway shields were taken down over thirty years ago. (CalTrans)

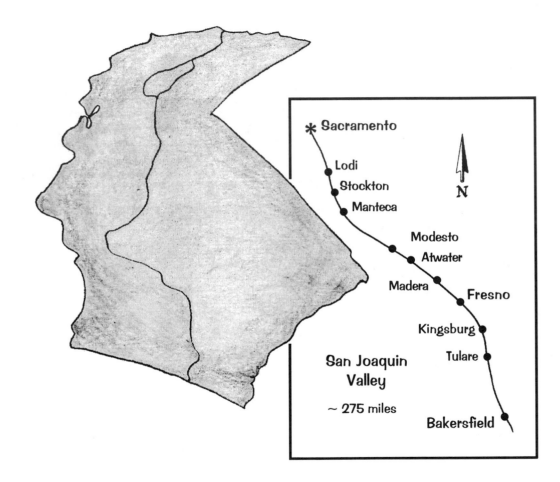

* Sacramento

Lodi
Stockton
Manteca

Modesto
Atwater

Madera
Fresno

Kingsburg
Tulare

San Joaquin
Valley

~ 275 miles

Bakersfield

N

14

CHAPTER

2

THROUGH THE GREAT LONG VALLEY

Our journey down Highway 99 to its end (or beginning) at the Mexican border starts alongside the State Capitol building. The old highway took us right into the heart of town, unlike today when the major routes invariably skirt the downtown areas. The Capitol grounds abound with plantings collected from the length and breadth of our long, great state. They are as pleasing today as when "gypsying" families picnicked and frolicked before seeking, guidebook in hand, a nice auto court in which to spend the night.

Perhaps, as we are doing, the family headed south on 99 toward Stockton. In those days, the bustle of the capital city was quickly relegated to the rear view mirror. As the auto camps, auto courts, and later, motels, that fringed the outside edge of town presented themselves, the vineyards and orchards and fields spread out into the leafy distance.

We continue to traverse the backbone of the Central Valley and of California itself. The Sacramento Valley to the north and the San Joaquin Valley to the south mesh seemlessly to form the 430 mile length, the great heartland of California. A huge chunk of land that cannot be ignored. Not large in population perhaps. But in sheer, flat land mass, in its verdant span of fields and orchards thriving beneath the hazy summer heat and winter fog... A force to be reckoned with.

Highway 99 bisected this valley from the dawn of the highway era in the first decade of the 20th century, and at the same time held it all together. The myriad little agricultural towns were united by the narrow ribbon of humble white cement.

As dirt roads evolved into paved highways, driving was elevated from the constant frustrations of digging out of mud or sand into something not only feasible but practical and, yes, fun. Traveling from town to town, for pleasure or for business, up to the Capital or down to Los Angeles, could now be done with a reasonable amount of time and effort, while in control at the wheel of a motor vehicle.

Railroad Origins

It is true that many of the Valley towns owe their existence to the railroads which arrived in the late 1800s. The chosen site for a train stop invariably led to the growth of at least a small town in that locality. But over time the trains, with their restrictive timetables and scheduled stops, grew wearisome. Americans, Californians, Central Valley residents, everyone longed to experience the boundless possibilities that the unfolding 20th century held in promise. We took to cars and roads and driving with unbridled enthusiasm, from the time in 1912 when that first single slab of 15 foot wide concrete was poured up to the present. We love (and need) our cars and our highways!

Dirt Track to State Highway

Much of the original alignment of Highway 99 is still intact going down through the Valley. Being so flat, obstacles to road construction were few. The first paved road was laid down with not much more than a little grading of the roadbed by a horse or mule drawn scraper. The occasional bridge had to be improved or built.

The haphazard mishmash of dirt roads that existed at the turn of the century started out as foot trails in the days of the padres. Those narrow paths widened out and became wagon roads with the crush of population brought on by the discovery of gold. And those wagon roads of necessity became automobile roads after cars appeared. The need for a system of serviceable roads, and the logical role for the government to play in building and maintaining roads, was recognized early on.

The concept of a State Highway System that would connect every county seat to the state's major cities via all-weather roads was embraced. Those early, heady days of the

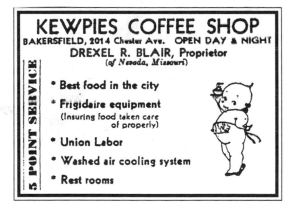
17

Looking south along a busy stretch of Highway 99 between Sacramento and Stockton near Florin Road in the 1950s. This highway remnant is now called Stockton Boulevard and shortly after this point is subsumed by the freeway version of 99 that opened in 1962. (CalTrans)

Highway Commission (which eventually evolved into CalTrans) were full of planning and dreaming and exploratory trips in buckboard wagons, with no funding available to actually begin constructing, improving, and paving.

Automobiles became ever cheaper due to assembly-line building and could be purchased by a new method called "installment buying". A series of Bond Issues was easily voted in starting in 1909. Existing roads were figured into the plan where possible, new ones planned where gaps existed, all to be brought up to the aspired State Highway standards... And thus we set out on a road-building trip that, almost ninety years later, has sometimes slowed but never stopped.

Cars were quickly deemed eminently desirable and soon thereafter, a necessity. As automobiles became commonplace, the highway through the flat valley was easily widened beyond the initial 15 foot wide, 4 inch thick strip to allow safer passing. Then another lane or two was added.

The concrete was then covered with asphalt. And, when traffic turned from a blessing to a problem, the old road was either covered over by a new expressway, or relegated to the status of a "business loop" or frontage road, which is much of what remains of old Highway 99 today.

Valley Towns

Yet always, and still, the old highway through the Central Valley held its string of charms. Certainly not pearls but possibly some unpolished gems: the small communities with interesting names dotting the length of Highway 99. Towns somewhat indistinguishable to the passersthrough, yet each to this day clinging to life, struggling to maintain an identity so easily quashed by the highspeed interstate whizzing by off to the side. Some have been more successful than others. Some were or are the "Capital" of this or that agricultural product, some built welcoming arches to greet travelers in the early part of the century, some have even played roles in recent history. The grape strike in Delano. The school bus kidnapping in Chowchilla.

Food, Fog, and Water

The present day traveler might view the Central Valley as an unwanted impediment postponing arrival in what is thought to be a more important place. Yet what is more important than food! And growing food is what the Central Valley is all about. That and water. Water in the form of irrigation projects that make all of the growing possible. For the Valley is relatively arid, and more so the further south you go.

And water in the form of floods is a recurring theme in Central Valley history. 1938, 1955, 1997. The low-lying valley is an old lake bed. Spaniards described the Valley in the mid-1800s as "a valley of lakes." When the rains do come, the water can't be held back. Watercourses, usually sucked dry for irrigation purposes, fill to overflowing. Levees break. Homes and fields and highways go under. It doesn't happen that often—we are in California—but often enough to remind us of who's really in charge here.

Pixley in Tulare County is one of the numerous small agricultural towns that dot Highway 99 in the Central Valley. The town once showed its civic pride by hanging the sign that spanned the highway, a common practice early in the century. Some of the old buildings and the row of trees along the train tracks are still there. (CalTrans)

Another grim reminder of who's boss is the dreaded tule fog. These days one might wonder, what is a tule, and what does that have to do with the Central Valley? Keep in mind that the Central Valley is the most human-manipulated environment in all of California. All for the sake of agriculture. The amount of food grown here is almost miraculous. But the natural face of the land has been seriously altered.

A hundred and fifty years ago there were about four million acres of marshland in the Valley full of reedlike plants commonly known as tules. 94% of that acreage is gone. All that remains of the tules is the tule fog, a dense ground-hugging fog that might not lift for weeks on end during the winter months. This dangerous fog, the cause of multi-car pileups, has always been the bane of highway travelers through the valley...

Our Humongous Garden

The forenoon traffic on the highway increased, salesmen in shiny coupes with the insignia of their companies painted on the doors, red and white gasoline trucks dragging clinking chains behind them, great square-doored vans from the wholesale grocery houses, delivering produce. The country was rich along the roadside. There were orchards, heavy leafed in their prime, and vineyards with the long green crawlers carpeting the ground between the rows. There were melon patches and grain fields. White houses stood in the greenery, roses growing over them. And the sun was gold and warm.
Grapes of Wrath, John Steinbeck, 1939

This is what the Joads saw in 1935 and the scene is much the same today, if not quite as idyllic. Some of the crops have changed. We now see exotic kiwis and pistachios. But fields of grapes and cotton still extend for miles.

It all started with the 1849 Gold Rush when some frustrated miners found their fortune by growing food for those with the picks and shovels up in the mountains. The railroad stimulated further development by providing a means to transport the crops to market when it came through in the 1890s.

During and soon after the Gold Rush era much of the Valley floor was planted in wheat. Water for towns and row crops came from artesian wells. But more was needed. It wasn't long before farmers banded together to form irrigation districts all up and down the Valley. They were charged with building dams and canals to bring water to the fields. More and a greater variety of crops could be grown, and quickly were. Not content to sit idle, it seems that westerners forever sought to expand, improve.

The federal government eventually came into the picture, initiating larger, more costly "reclamation" projects. Every creek and river was exploited as it came out of the mountains which frame the Valley on the east. The water was sent down into the Valley via watercourses of the farmers' choosing. Never was this push for more considered ecologically unsound. That concept wasn't even invented yet. The

Carrots mature in a field , a water tower marks the Weed Patch migrant labor camp mentioned by John Steinbeck in Grapes of Wrath. *The camp, now populated by Hispanics rather than Okies, is located a few miles off of the old highway but is definitely a part of the 99 story.*

destruction of the natural environment wasn't lamented. It was our right, even our duty, to bring this dormant landscape into the most productive cultivation.

These were the forces that shaped the Central Valley in this century. That gave birth to the extensive fields and the little towns. That brought the Valley to the point that now, a quarter of the food eaten in the United States in a given year is grown here. The Central Valley is an impressively vast and varied garden, sustained by irrigation ditches and immigrant labor, connected to the rest of the world by the length of pavement known as Highway 99.

Pick, But Don't Eat

Of course, appearances aside, it didn't prove so idyllic in the Central Valley for the Joads. *The Grapes of Wrath* dramatizes the dilemma of going hungry in this land of plenty, the family burning up precious gasoline traveling up and down Highway 99 in their ramshackle car seeking work.

The cornucopia of the Central Valley could not be filled without workers, and plenty of them, willing to work hard and travel from town to town. Traditionally most of these workers have been immigrants. From Oklahoma during the Depression, generally from Mexico in later years. The Valley farms provided needed work for displaced people, hard work that more established people didn't want to do. Enough work to survive on; not enough to thrive on. The Valley sustained itself and grew only because of the influx of agricultural workers. The reverse could also be said to be true.

The Central Valley is flavored by this reality. The guidebook *California, A Guide to the Golden State*, one volume of the American Guide Series written by the Federal Writer's Project during the Depression years, had this to say:

...The migratory worker, constantly on the move to catch the harvest seasons of one crop after the other—peaches, walnuts, apricots, grapes, celery—never stays long enough in any area to establish himself as a citizen. He lives apart from other residents, occasionally in barracks behind the fields and orchards, more often in crude shelters of his own devising along the river bottoms. Because there are too many who want to work, the migrant cannot command an adequate return for his labor. The inhabitants of the towns do not know him and his family and local governments feel no responsibility for him. No one knows how to help him and no one knows how to get along without his help.

Upper Valley

From the capital of Sacramento to Bakersfield at the south end of the Valley the distance is about 280 miles. This portion of Highway 99, in the old tradition of naming highways, was known as the Golden State Highway, or the Valley Route. We will use *California, A Guide to the Golden State* as our guide through the Valley, and all the way to end of the road. Using this book sixty years after it was compiled gives a real sense of "how it used to be." Some earlier guidebooks were based on passenger train routes, but by the 1930s motoring was the preferred mode of travel.

The small towns started to appear not long after leaving the State Capital by heading south on Stockton Boulevard, Highway 99.

Galt, Lodi, Stockton

The web of highways spreading across the land often combines and converges near the larger cities. For the approximate fifty miles from Sacramento down to a little below Stockton, 99 originally shared its routing with a worthwhile companion, the Lincoln Highway. While Highway 99 crossed the west south to north from the Mexican border to the Canadian border, the Lincoln Highway (most of which became Highway 50 in California) was the very first established east-coast-to-west-coast cross-country highway.

In contrast to Highway 99, which was already an established route that was later given an official highway designation, the Lincoln Highway (as were many of the named highways of the teens) started out as a concept only, its actual course yet to be determined by its promoters.

Unique neon motel signs seem to be disappearing, but some nice examples are left such as this one in Stockton.

27

For these early named highways, generally the most direct route was sought. But on the other hand, towns offering to share the cost of construction were often picked to be put on the highway route. The route was then publicized, mapped, and signed by the various highway associations, all to the benefit of towns along the way. Possibly that is why the Lincoln Highway doglegged in a straight southerly direction after reaching Sacramento, sharing the road with 99, before it turned back west to San Francisco (its termination point) near Stockton.

Galt...*produces poultry and dairy products, figs and grain. Lodi...is a typical valley town: the streets almost all wide, straight, and lined with trees that give a deceptive promise of protection from the intense heat in summer. Proud of its vineyards, it holds a Grape and Wine Festival in September.* A forty foot wide lath and plaster arch was built in Lodi in 1907 for the first such festival, then called the Tokay Carnival. It still spans Pine Street, a couple of blocks off of old 99, and is thought to be the oldest surviving welcome arch in America.

Stockton became a large and important town because of it's distinction as a sea port, even though situated up a river sixty miles inland. On leaving Stockton, *US 99 moves south across a flat expanse covered with great walnut orchards, truck farms, melon tracts, vineyards, and pastures.*

For a short period of time, two

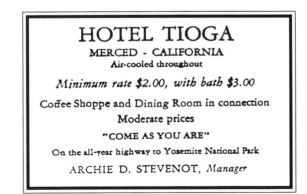

Highway 99s left Stockton heading south, a 99E and a 99W. That split lasted only five years, from 1929 through 1934. The short lived 99W continued south of Stockton along the Lincoln Highway route for a few more miles (basically where I-5 runs today) then came back across Yosemite Avenue to converge with 99E at Manteca. 99E followed the route (now covered) that was Highway 99 from 1909 until the freeway was built in 1955.

Manteca

The ignoble but humorous name of the town, which translates from Spanish as "lard," was acquired by accident. The town at its inception was a railroad stop called Montcca. "Mantcca" was accidentally printed on the first batch of train tickets, a mistake that was never corrected.

Until recent years, you always knew you were approaching Manteca by, dare I say, its odor. That strong smell, somewhat offensive to outsiders but hardly noticed by residents I would guess, was the smell of success, of two prosperous local industries that kept the town going. The town housed a feed lot, and "Manteca-fed beef" was famous around the state (not for the smell but for the taste and tenderness!) for some time. What the beef were fed was the stinking fermenting residue of sugar beet

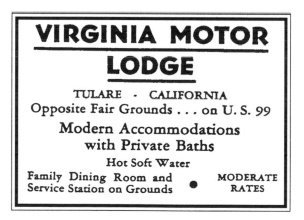

processing produced at the Spreckles Sugar plant next door. Both plants now stand empty and unused alongside Highway 99.

Manteca was also an important highway crossroads. Yosemite Avenue was heavily traveled by people from the San Francisco Bay Area on their way to Yosemite National Park, a favored destination from the time that car camping became the craze in the teens. Manteca was a good point for a rest stop, and the usual stop was at the Manteca Creamery for an ice cream. The intersection of 99/Yosemite Avenue in downtown Manteca became a troublesome bottleneck that was not eliminated until the freeway bypassed the town.

Ripon, Salida, Modesto

Around Ripon radiate large vineyards....A large roadsign proclaims the sprawling, white, peak-roofed building as the "World's Largest Exclusive Brandy Distillery." ...Ripon also claims "The World's Largest Out-door Rummy Game." One wonders just how much competition was involved to win that distinction! *Salida, a fruit, alfalfa, and grain shipping point, is a center for dove and quail hunting*

In Modesto, *the streets, crossing at right angles, run northeast, southwest, northwest, and southeast to give the maximum of shade during the days of intense heat.*

Oleander plantings in the median strip (here north of Modesto) were more than ornamentation. The tall bushes were a safety feature that blocked the glare of oncoming headlights. (CalTrans)

99 OFFSPRING

Highway 99 as we knew it in the 30s, 40s, and 50s became official in 1926, at the time that United States Highways were assigned their numbers. US Highways were selected portions of the State Highway system that were partially funded by the federal government. So the road known in various segments as the Golden State Highway, the Valley Route, the Ridge Route, and probably other localized names that have fallen by the wayside, and that had been identified by a confusing array of numbers imposed by the state of California, were all linked by the white 99 shields erected all the way from Blaine, Washington to Calexico, California.

Three other federally funded highways branched off 99 in California and in a sense were its off-spring. Highways 199, 299, and 399 were east-west "laterals" that connected north-south Highway 99 to one of the other major north-south California highways, the coastal Highway 101 or east side 395.

There were many east-west highways but we know these in particular were considered offshoots of Highway 99 by the inclusion of "99" in their numbers. The standard numbering convention designated east-west highways as even-numbered. Those federal-aid highways designated as connecting laterals retained the major roads' number (99) with a third number added to the front.

Of the three 99 offspring, two retain their original numbers. Farthest north, 199 goes from 101 at Crescent City on the far north coast to 99 at Grants Pass, Oregon and is still a US Highway. 299 kept its number but was decommissioned and relinquished to the state in 1964. It starts on 101 in Arcata, crosses 99 at Redding, and continues east to 395 at Alturas.

Much farther south, in the territory covered by this book, what was 399 (before it too was decommissioned) linked Ventura on 101 to Bakersfield on 99. This is still the main road connecting the Santa Barbara area to the San Joaquin Valley but now goes by the numbers 33 (Ventura to Taft) and 119 (Taft to Bakersfield).

*Modesto has canneries, dairies, packing plants, and warehouses...*The town also has a welcome arch, erected in 1912. It stands on I Street alongside the old highway bearing a slogan that is indicative of the local obsession with water. "Water, Wealth, Contentment, Health" it says. The author of the motto was awarded the $3 Second Place prize in a contest. The first choice, "Nobody's Got Modesto's Goat," fell out of favor before the arch even went up. Modesto retains the small town flavor that inspired the 1973 movie *American Graffiti,* written and directed by George Lucas.

Mid Valley

Ceres, Turlock, Livingston

In the neighborhood of Ceres, checkered with vineyards and fig, peach, apricot, and pear orchards, the date gardens are of particular interest...US 99 crosses and recrosses a network of canals on its way to Turlock, in the midst of hundreds of small farms...The town celebrates its watermelon crop at an August watermelon carnival.

Livingston, trade center of a sweet potato belt, also ships raisins, grapes, peaches, and alfalfa. More recently, Livingston has been known as home of the last surviving traffic signal on Highway 99. The period of being glad that motorists were forced to stop in town was decades past when the freeway bypass was finally completed in 1997. The dangerous, unexpected intersection was an infamous "blood alley" especially during thick winter fogs. A landmark diner, the Blueberry Hill Cafe, was torn down during the construction, and the Foster Farms restaurant alongside the chicken processing plant which stands near where the signal made its last stand is now vacant.

Atwater, Merced, Chowchilla, Madera, Herndon

Atwater styles itself the "Home of the Merced Sweet Potato"....Merced...is the principal rail and motor gateway to Yosemite National Park. It is in a grazing and hay and cotton-producing area and has a cotton gin, cement factories and potteries...

Chowchilla is in a district of dairying, hog and poultry raising, cotton, fruit and grain growing....The Chowchilla River is referred to locally as the region's Mason and Dixon Line; legend has it that Union soldiers marching south from Stockton during the Civil War were ordered to load their guns when they reached it.

Just south of Chowchilla is the only remaining, still operating "giant orange" stand on this stretch of Highway 99. Giant Oranges and their knockoffs (Mammoth Orange, Big Orange, etc.) proliferated along California highways, and Highway 99 especially, from the 1930s to the 1950s. Sacramento, Galt, Lodi, Turlock, Merced, Chowchilla, Madera, Fresno, Tulare, and Bakersfield all were home to one of the Giant Orange chain, with the imitations scattered inbetween. After all, you could build a stuccoed ball for about $1000, paint it orange, and the motoring families would be irresistibly drawn in for a cool drink and a snack in those pre-air conditioning, pre-McDonalds days of old.

Madera...produces great quantities of sweet wine...South of Madera the grape vineyards are displaced by cotton fields and orchards. The small cluster of frame houses in Herndon is bowered in fig orchards, which press to the edges of the highway and railroad tracks.

Also a short distance south of Madera, a significant landmark remains, trapped in the median strip between the racing traffic of the current incarnation of north

and south bound Highway 99. Look quickly and you will see a pine tree, representing northern California, and a palm tree, representing southern California, standing side by side at what was approximately the half way point of Highway 99 when it ran the whole length of the state.

South Valley

Fresno, Fowler, Selma, Kingsburg, Goshen

By the early 30s, most of the length of 99 through the Valley was widened to at least twenty feet. The portion south of Fresno was widened to thirty feet to accommodate the Valley's heaviest agricultural traffic.

A six mile section north of Fresno was an early site of an ill-conceived traffic-relief idea, a middle passing lane to be used by both directions of traffic. The idea was put to use along other parts of the highway as well, but the third lane quickly became known as a "suicide lane," and maybe was not such a good idea after all. It was phased out when four lanes became the norm, an improvement that had been delayed by World War II.

Fresno, with its many well preserved historical buildings and several ornate movie theaters, retains some of the atmosphere that greeted the early to mid-century Highway 99 traveler passing through this small city. A welcome arch erected in 1925 spans what once was Highway 99, the main artery into town. The arch stands in what is now a seldom visited part of town but it still proclaims Fresno to be

The Fresno Arch was erected in 1925 and refurbished in the 1970s. The arch is now seldom seen by visitors for what was once the city's main entrance is now a quiet warehouse district.

"The Best Little City in the USA." The surrounding area abounds with vineyards that produce the raisins that made the county famous.

Fowler is an important horse and mule market. It has several whisky warehouses, grape processing plants, and fruit packing houses. Selma produced raisins *that come from the vine in large, full clusters...dried on the stem for the fancy Christmas trade.*

In the foothills around Kingsburg are many peach orchards, vineyards, and orange groves. The population, still 90 percent Swedish in descent, is made up of the children and grandchildren of a colony of Michigan Swedes that settled here in the 1870s. Kingsburg today houses the Sun Maid Raisin plant, and capitalizes on its Swedish descent. Near Goshen, *the 7,000 acre Taugus Ranch advertises itself as the "World's Largest Peach, Apricot, and Nectarine Orchard."*

Tulare, Tipton, Earlimart, Delano, McFarland, Famoso

The Hotel Tulare has more than 1,000 mounted native birds and mammals, and an extensive collection of California wildflowers.....Fields of cotton, white-tufted in the fall, border US 99 on its way to Tipton...At Earlimart the one-story frame-brick buildings are dwarfed by the huge cotton-oil plant, the great cotton gin, and the long sheep and wool loading sheds.

Delano was put on the map in 1965 as the headquarters of the United Farm Workers. Years of unrest among the minority farm workers were brought into focus by the leadership of Caesar Chavez. There had been small uprisings in the previous thirty years. But the farm workers who were poor, with no real home, and speaking

English as a second language if at all, had been more concerned with mere survival than with the high-minded ideals of improving the lot of all farm workers. The grape-buying boycott and the strike on many of the large-scale grape growers in the San Joaquin Valley concerned much more than higher wages. The ultimate goal was union representation and improved living and working conditions. Buoyed by the national mood that had been fueled by the civil rights movement and the determined persistency of the union organizers and followers, the strike and boycott were successful.

McFarland *drew for its first settlers people opposed to the whiskey drinking tolerated in nearby Delano and Famoso. All land deeds contained a clause prohibiting the selling of liquor...today McFarland's bartenders do a lusty, legal business.*

Bakersfield

Approaching Bakersfield, *the massed rigs of the Kern Front oil field stand beyond a strip of fruit orchards.* Here, at the narrower south end of the great valley, the landscape changes, a welcome variation to travelers motoring down the Golden State Highway, Highway 99. Oil fields mingle with crop fields. One percent of today's world oil production comes from this area. The Tehachapi Mountains, round and brown in late summer, cloaked in green grass and an unbelievable carpet of California poppies in the spring, lurk down the road in the distance.

In the Bakersfield of 1939, *the downtown district is metropolitan in its variety of shops, cafes, department stores, theaters, and office buildings. In its streets great motor vans of potatoes, lettuce, and grapes are as familiar a sight as trucks loaded with oil well casing, drilling equipment, and derrick parts.*

The old highway route, on its last leg of the journey through the Valley, leaves Bakersfield via a long ash and palm tree-lined boulevard, oleanders in the median. *US 99 shoots south across the floor of the San Joaquin Valley in a course that bends only once.*

Ahead lies...relief from the unrelenting flatness. A mountain barrier. A formidable hurdle to be crossed as we continue our trek to southern California.

Tomato Tie-up:
Truck Loses Load in Highway 99 Crash

Motorists on southbound Highway 99 found their cars up to the wheel wells in a crude salsa Tuesday morning after a tomato truck tangled with three cars and spilled its load. About 50,000 pounds of Roma tomatoes closed two lanes of the highway for about three hours.

The 9:30 a.m. accident caused one minor injury, California Highway Patrol Officer Joseph Frasier said. The tractor pulling two trailers was traveling in the third lane from the left when traffic stopped at the 12th Ave. offramp. The truck was unable to come to a stop in time and caused a chain-reaction accident involving three other vehicles.

Motorist Jim Hare was on his way to pay tuition at his child's school when his new Ford Ranger pickup was struck from behind by the rig.

The force of the crash bent the truck in half, but Hare escaped injury.

Hare was upset over losing his truck, but he managed to find humor in the situation.

"This is hilarious," he said, watching as television cameras circled around his car. "Just what I've always wanted--to be on the news."

Caltrans used a scoop to get the tomatoes off of the road. The cleanup took about six hours.

By Ralph Montano © *The Sacramento Bee* 8/19/98

39

 SIGNAGE CALEXICO ⇨

We take our road signs for granted. What road we're on, how far to the next town, all displayed on uniform signs erected by the highway department. It wasn't always the case.

If the earliest automobile roads were signed at all it was a home done effort; a painted board nailed to a tree advertising the garage or hotel in the next town. The earliest motorists confronted a confusing array of dirt roads and unidentified crossroads. The smart motorist kept a road guide in the glove box to help make sense of it all.

Thorpe's Illustrated Road Map and Tour Book, 1911 covered all of California (including what became Highway 99) mile by mile with a profusion of details. "Follow bottom of hills." "House and eucalyptus tree here." "Follow river bed 3.2M." "Palm tree at turn."

The auto clubs took the sign matter into their own hands early on. Automobile clubs emerged soon after autos themselves early in the century. They were a part of the widespread Good Roads movement that successfully lobbied for more and better roads.

Starting as early as 1906 the Automobile Club of Southern California began erecting their porcelain enamel on 18 gauge steel road signs, which included their own logo, on southern California roads. By 1910, 2,400 signs had been installed. The California State Automobile Association did the same for the central and northern parts of the state beginning in 1914, with 84,000 erected by 1928. The clubs (and their dues paying members) paid all costs until the cities, counties, and state agreed to kick in some money (for signs only, no labor or maintenance costs) in 1933.

In 1926 the U.S. routes were designated. The 99 shields went up, thanks again to the auto clubs. The Highway 99 signs of the 1920s through 1930s had squared numbers, as did all of the U.S. route signs. The rounded numbers of the 40s and 50s are more familiar to most of us. Glass reflectors were replaced with plastic during WWII and new signs erected for the duration of the war were made of wood. The clubs' signing duties did not end until 1956 in southern California and 1969 in northern and central California.

Where Will We Sleep Tonight?

Early motorists were a rugged lot who took to the road with an adventuresome spirit. Back then, automobiles were viewed not only as a mode of transportation; they were the embodiment of a new kind of freedom. No particular place to go. No restrictive schedules. A careless meander along a narrow road. Or, a mild challenge: man and car against mud, sand, or mountain.

It was slow going in the first decade of the 20th century. Motoring more than a few miles from home would inevitably require an overnight stay someplace. But where?

Many early motorists hated hotels. Too stuffy. Too urban. Too expensive. Dress codes. Bland food. Tipping. It's true that many motor travelers did indeed put up with the snootiness and restrictions of hotels. Driving could be a rough and dirty experience. A hotel's comfort at the end of the day could be worth the necessity of arriving early to secure a room, ignoring the stares at rumpled, informal clothing, or

waiting impatiently for the garage to open before the morning departure.

The logical solution was to camp. If some motorists did not actually practice it, they surely embraced the idea, at least until something better came along. Roadside camping fit in well with the whole freedom-of-the-road concept.

The car-camping motorists considered themselves to be a special breed and called themselves "gypsies." They were a vocal minority. A vast array of camping equipment; fold out beds and tables, thick canvas tents, nesting cookware, soon was touted to the driving public.

But the towns along the highways were loathe to lose the "tourist dollar" that could be

spent, if not at the hotel, then in the cafe, gas station or grocery store. By the early 1920s, small towns and large cities alike provided campgrounds in a city park or at the edge of town. Tables, showers, gas stoves, electric lights and wash tubs, all for free, and very popular.

By the end of that decade, the municipalities had lost some of their benevolence in the face of transients and other "undesirables" taking up semi-permanent residence in the camps. The towns started charging fees and implementing time limits. By the early 30s, most of the municipal camps were closed.

Private camps took up the slack, on the fringes of town or farther out in the country. These entrepreneurs were quick to expand on the amenities offered to the motorists, and the motorists took to them wholeheartedly. A farmer's roadside fruit stand expanded to offer camping facilities. A small cafe was built. Flimsy walls and a roof were hammered on to tent platforms, and primitive cabins were now for rent. But don't forget to bring your own bedding and frying pan, and the bathroom is communal! Tent sales plummeted.

That was the humble beginning of motels. The earliest motels, as "auto camps," rented both campsites and simple cabins. "Auto courts" were the next stage of development. These more comfortable cabins (bedding and lavatory now provided) were joined in rows, each separated by an open car port. Leave all of the camping gear at home,

---

Camping in Comfort

...Every camper owes it to his own manhood as well as to his fellow beings to keep his camp clean and sanitary while occupying the site, for a night or for a month. To mar the sweet cleanliness of nature, to spoil the purity of a forest or the brightness of a grassy slope of a hillside is a sacrilege and utterly inexcusable. Camp should be cleaned every morning by burying all garbage refuse, to the extent of burning out empty tins and burying them with all other unburnable trash. In more permanent camps dig a trench or sink a hole and burn out all refuse matter with a brush fire or some kerosene. This procedure will keep flies from getting at it and carrying contamination far and wide. All organic refuse should be covered with earth or buried instantly. Leave every camp site as clean or cleaner than you found it, all ready for the next occupant...
Motor, June, 1920

The earliest motorists often camped. Hotels were pricey, formal, and located in towns. Auto courts and motels were in infancy. Staying in established auto camps, or just plain camping out, was favored by these self-described "gypsies."

and still no tipping or dress codes.

"Auto courts" became the "motels" of the 40s and 50s. Motels flowered in that era, all independently owned. They were constructed in a variety of building styles (wig-wam village to Spanish mission to simpler U-shaped court) and marked by distinctive neon signs.

Today, the distinction between motels and hotels at one extreme, and motels and cheap by-the-month apartments at the other extreme, is vague. It seems that hotels are either very upscale or very downscale, and motels are everything inbetween.

Stacy Vellas Remembers Highway 99
Her Father, Family, and the "Fruit Tramp Trail"

My Dad was from the Old Country and he had his quaint expressions. He believed in the whole family working. There was no feigning illness to get out of work. But he'd always set a goal so we knew when we would be finished. He inspired us by setting up races; who could pick the most bushels, the most hampers.

When we'd pick potatoes he'd say, "We pick 200 bushels, we go home." We worked in pairs constantly racing to see who could do the most. He tried to instill in us that work was something to be enjoyed. When we worked "piece work", the faster we worked the more money we made.

He was very tight with money, but he always gave us money to go to the show, every night if we wanted to; and he always took us for a milkshake at the creamery after work. He loved getting us a treat. He was a neat Daddy, a very special person.

Old Highway "99" - The Trail of the Migrant Worker   by Stacy Vellas ©

I turned sixteen that year
When Daddy bought the yellow bus
And said, "We're going to California
To make a new life for us.

California's the promised land
There's lots of work I'm told.
We go; we make our fortune
A better life my kids will know."

We pitched our camp near Redlands
But rain had closed the sheds in town.
"There's work in Imperial Valley," they said
So we packed and headed on down.

The pea fields stretched along for miles
Near Calipat in 'forty three.
And little kids, too, carried hampers
And picked the sweet green pea.

The man at the scales paid on the spot
And the coins filled up my pocket.
The sweet fresh smell of the cool, damp earth
And the money, we couldn't knock it.

A tent house in a Government Camp
Mama made into a home.
And that old yellow '29 school bus
Carried everything we owned.

By summer we'd crossed the Grapevine
Down into the San Joaquin.
From Bakersfield to Penryn
Our "mobile home" was seen.

We camped along the highways
And slept in the back of the bus.
Daddy built a fire 'long side the road
And Mama cooked for us.

The next Government Camp was our new home
When the fruit hung ripe on the vine
And Old "99" was the Fruit Tramp Trail
With a new job down the line.

But the open road was a heavy load
For my sisters in grammar school.
For kids who missed three months a year
Had lots of catching up to do.

Till Daddy bought a farm and settled down
And the bus came and took us to school.
He raised his crop on a five-acre plot
And peddled the things he grew.

Daddy always said to me,
"Ray, girl, it's good to work.
It makes the mind and body strong
And the money doesn't hurt."

45

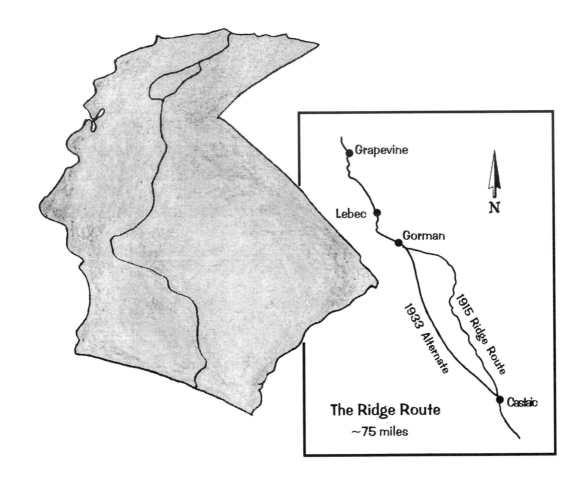

The Ridge Route
~75 miles

Grapevine

Lebec

Gorman

1933 Alternate

1915 Ridge Route

Castaic

N

CHAPTER
3

BREACHING THE
BARRIER BETWEEN NORTH AND SOUTH

"The Ridge Route." "The Grapevine." This portion of US Highway 99 that crossed the mountains between Los Angeles and the Central Valley was the most challenging segment of the road to build between Sacramento to Calexico, and could as well be the most difficult to travel. But more important to note is the significance of this new link between the north and the south of California that was forged when the highway first opened in 1915.

The sight of the Tehachapis is a welcome one after the unrelentingly flat journey down the Valley. These deceiving mountains, looking smooth and round from the Valley approach, quite effectively blocked easy travel between southern

California and the rest of the state in the early days. Even now it can be a difficult or impossible crossing when severe weather takes complacent California drivers by surprise.

Usually, it is a fairly quick, if decidedly steep, 40 mile hop up over the mountains from Grapevine (on the north side) to Castaic (on the south side) on I-5. Throw in a winter storm, a truck accident, a holiday weekend and the passage is an often discussed real life impediment to speedy, troublefree travel.

Undertaking the mountain crossing via "The Ridge Route" or "The Grapevine," as this part of what was Highway 99 is often called, still presents a mild challenge and seems to encompass the allure of a piece of terrain to be conquered.

Interestingly, both of these common names are misnomers. The actual "Ridge Route," the original State Highway through these parts that did indeed take the ridges, is way off to the east of the freeway and has not been in use since 1933. "The Grapevine" technically refers only to the portion of the highway as it winds down Grapevine Canyon on the approach to the San Joaquin Valley. The road itself used to twist like a grapevine, but the name, bestowed on the creek and canyon by Spaniards in the 1800s, comes from the wild grapes that grow in the region.

Highway Relics Today

The mountainous portions of highways seem to have had the most extensive upgrading and realignment over time, as earth-moving and bridge-building technology advanced and public demand for better roads in the form of easy grades and wide curves increased. Over the Tehachapis there have been four versions of Highway 99, (the last one being I-5) and there remains at least one location where pieces of all four can be viewed from one spot. When a new, straighter version came along, the former road was chopped into bits and pieces that take some detective work to reassemble. Some pieces were abandoned altogether, others turned into local or parallel frontage roads, others covered up by the new road. Today's I-5 in different places encompasses portions of all three previous highway alignments through here.

The first State Highway, officially called the "Tejon-Castaic Ridge Road"

Looking south up Grapevine Canyon in 1958, the 1915 Grapevine Grade twists its way down the left side of the canyon, while the road opened on Sept. 11, 1934 comes down the right side. Currently, southbound I-5 follows the later road. Northbound I-5 cuts across the curves of the early road, leaving a few isolated fragments for us to find. (CalTrans)

but known far and wide as the "Ridge Route," was partially open but incomplete in 1914. The grading and oiling was completed and the highway officially opened in 1915, but it wasn't paved until 1919.

This road was already obsolete, though still in use, by the time it became a Federal Highway and was given the "99" designation in 1926. When the new road was built, that first State Highway was left largely intact since the next, three-lane version, the Ridge Route Alternate, was built some distance off to the west roughly where I-5 runs now. It opened in 1933.

The cumbersome word "Alternate" fell out of use, and more lanes were added as needs increased. Most of the Ridge Route Alternate was upgraded to a four-lane divided expressway by 1952. By the late sixties, Highway 99 over the Tehachapis was no more. The "Ridge Route" is now comprised of the eight lanes of I-5. In various places across the mountains, remnant pieces of the three previous versions wait to be discovered.

A Unifying Force

The Ridge Route [is] a great and powerful influence in promoting the unity and integrity of heretofore divided sections of the State and in discouraging state division agitation. The new road entirely does away with the old 30 per cent grades which took the stamina out of a motor on the Midway Route. (California Highway Bulletin, 1916)

Across this twisting alignment, an imaginary division extended to separate the great box valley of California from the south, a division which seemed to have been recognized by

Nature for, to the south, everything was suggestive of the desert—bald hills, sage brush and yucca. On to the north as the divide was crossed, small and stunted trees began to appear and developed into such forestation as marks the "Sanctuary of the Pines" on Frazier Mountain...(National Motorist, 1933)

Try to picture the state at the turn of the century. The estrangement of northern Californians from southern Californians was extreme. Talk of splitting the state in two is not just a modern idea. Communications were agonizingly slow.

In 1910, motoring from Los Angeles to Bakersfield required a circuitous 170 mile day-long (if you were lucky) routing through the Mojave desert, known as the Midway Route, to avoid the ruggedness of the Tehachapis. In 1912, bond money in pocket, the State Highway Commission set out to explore the most practical Los Angeles-to-Central Valley route. As was their mission, the most direct route within the building capabilities of the times was sought.

Chopping their way through the brush, clinging to the precipitate walls of canyons where no pack mule could keep his feet, across ravines and along the crests of the mountains, the surveyors fixed their stakes, and, link by link, laid the lines along which this mighty highway should run. (California Highway Bulletin, unknown date)

Unfortunately for them, a number of desert business men assumed that the old Mojave route would be chosen for the State Highway and had subsequently bought up adjacent land parcels. But the hardy Highway Commission survey crews had discovered a more direct route that shaved 45 miles off of the distance, was indeed feasible to build, and with grades no steeper than 6%. Nevertheless, the ensuing political battle held up construction for several months.

Soon enough, the workers hacked their way through the mountains, and the Ridge Route was built. It was hailed as a fantastic engineering feat, even with the narrow twists and turns that totaled 110 complete circles and kept the speed limit at 15 mph. Newspapers and magazines sung its praises.

Still, the road frightened travelers and challenged truckers. The views from

It was a steep and scary ride down the Grapevine Grade in the early 1930s, although these motorists are past the worst as they near the bottom of the grade and approach the Central Valley. The new Grapevine road completed in 1936 was an improvement, but to this day a vehicle's brakes are put to a harsh test along this stretch of highway. (CalTrans)

the tops of the ridges were spectacular, the driving dangerous. It quickly became the most heavily traveled highway in the west, and possibly the most hated. Awe turned to impatience. Less than twenty years old, it was replaced by the Ridge Route Alternate.

That it [the Ridge Route] *will always carry travel, in spite of the new, high speed way* [the Ridge Route Alternate], *is a foregone conclusion, for it will always be of scenic interest. ...A thousand vagaries of view will always attract the lover of nature, while the new road will appeal to those who place scenery second to speed. National Motorist, 1931*

The magazine was wrong in its prediction. It appears that just about everyone placed the value of speed above that of the scenery. In a matter of months, most of the string of businesses, the ridge-top hotels, gas stations, and eateries that lined the first highway, were abandoned.

Men, Mules, and Power Shovels

Let's go back to the time when Californians were marveling at the vast amounts of dirt being moved in the mountains, the twenty foot wide strip of cement being poured, and the number of men and machines it took to accomplish this, to forge this link through the mountain range that separated north and south.

The big push was to have the highway complete in time for the 1915 Panama-Pacific International Exposition in San Francisco that was to showcase all that California had to offer. Open it did, but only for the last few weeks of the World's Fair and only as an oiled roadway. Paving was delayed and could not be completed

53

until after the Great War.

The construction troubles began even before the climb into the mountains south of Bakersfield. The route from Bakersfield south into the mouth of Grapevine Canyon was barred by a five mile stretch of adobe and alkali soil stretching so far to the east and west that a detour was deemed too expensive. In wet weather *it became so gluey that a strong horse could not drag a light buggy through it* (*California Highway Bulletin*, 1916.)

Crews working on this "Grapevine Unit" of the new highway met the obstacle by building up a highway grade with hauled-in material. A specially constructed railroad (which occasionally lost a length or two of track in the mud) transported the material to the site. That remarkably straight stretch of highway was dubbed the 17-Mile Tangent.

At the end of the Tangent, the town of Grapevine was the last outpost before ascending the grade. All that remains of the original Grapevine is the shell of a small block of tourist cabins. The sleepy relic sits in a wedge of land between the north and southbound lanes of I-5, right at the foot of where the coils and spirals of the old Ridge Route began their steep ascension.

The flood-prone canyon bottom and frequent creek crossings that had plagued the rough old wagon road through here were avoided by hacking the highway roadbed out of the hillside well above the level of all recorded floods. Still (records having been kept only a fraction of a second in geologic time), high water washed it all away in one day in March, 1914. A higher hillside line was laid out, the roadbed rebuilt and this time it stayed in place.

So in a spectacular series of hairpin turns, the highway gained its elevation up out of Grapevine Canyon. One of the most infamous places quickly acquired the name "Deadman's Curve." The hillside below it was known as the Junkyard, no explanation required. That once dreaded fragment of the old highway rests, powerless and unnoticed these days, off the edge of southbound I-5.

After going up "the Grapevine," the highway climbed more gently on its sidehill path through a wooded valley, passed Ft. Tejon, then Lebec. The Lebec Lodge, a spacious two story hotel with lush gardens, became a favorite of tourists as well as the Hollywood crowd, being a couple of cuts above the standard tourist cabins of the day. Then the road topped Tejon Pass, and went on to Gorman. Near Gorman began the Ridge Route proper, a route still easily traced in its entirety. Of this 36 mile section, 29 miles actually does top a series of ridges.

Constructing the Ridge Route posed new challenges to the road builders, whose previous mountain highway experiences were limited to building roads up canyons. Using new methods, the Tehachapis were breached with sweeping banked turns and deep cuts, a system inspired by European alpine roads. The trade journals raved

"Fag Stations" Established for Smokers in Forests

Smoking has been prohibited in all National forests in California except at "fag stations" established along trails and roads and at camps and places of habitation, according to decree just issued by Regional Forester S.B. Show. A report states that 999 fires in 1932 were caused by lighted matches and burning tobacco being thrown into dry litter of the forest and dry grass and grain fields of the State.
California Highways and Public Works, Oct/Nov 1933.

Forest fires were a constant worry, especially along the Ridge Route in its course through the forest lands, but tobacco was not always the culprit. In September, 1928 an overturned truck sparked a fire that spread to 35,000 acres.

about the quantity of "material" moved, although initially the minimum amount of cutting and filling was done to keep the gradient at 6% or less. ...*Few indeed are the cars which cannot make the drive in high gear. (California Highway Bulletin,* 1916) The number and sharpness of the curves seems to have been of lesser concern.

110' deep Swede's cut, or the Big Cut, was part of the Culebra Excavations on the original 1915 Ridge Route. This was the only place in which rare and expensive steam shovels rather than men and mule teams were used for moving material.

The vast majority of the roadbed was roughly leveled by the 300 to 400 crewmen with picks and shovels. The roadbed was then graded by mules hitched to large iron bars known as Fresno scrapers. There were 250 mule teams on the job.

Work was hampered by the remote

location and the sometimes severe weather. Supplies had to be hauled all the way from Lancaster by mule teams or the fleet of sixteen trucks. Workers lived at the Liebre Road Camp, a site along the route, and often quit their arduous jobs.

Expensive motorized equipment was used for only the heaviest excavations. The few deep cuts ran up the construction costs but greatly shortened the mileage. Steam powered shovels were put to use on what was known as the Culebra Excavations. The largest feature was the 110 foot deep Swede's Cut, or the Big Cut. Two-shovel teams worked 24 hours a day. Chain-driven dump trucks with solid rubber tires hauled away the material.

With the initial construction done, the road was opened to an anxious public in October, 1915. A number of Ridge Route businesses sprang to brief life. The auto clubs installed their road signs (a duty they willingly performed for decades until the state finally took over) and marveled about the scenery and ease of travel along the Ridge Route in their publications.

The trip was not always as easy as had been portrayed. The traveler's needs; food, shelter, fuel, repairs; were met by a number of establishments. Traces of them are still found along the way, although it's been sixty years since the motorists quit coming. Sandburgs Summit Hotel (famous for Mrs. Sandburg's apple pies!) The Tumble Inn. The National Forest Inn. Reservoir Summit. The View Service Station. The Ridge Road House. Martin's. The Half Way Inn. Some were large timbered lodges. Most were more humble. In some, business was very good during Prohibition for obvious reasons made possible by the out of the way locations. It is said that (paid) ladies were also available for entertainment in certain places along the Ridge Route.

The auto club signs proliferated, especially the ones saying "Caution, Grades and Curves," "Blow Your Horn!," and "Warning, Motor Officers On This Road! Obey The Law! 15 m.p.h. Speed Limit!" "More signs than trees" as one auto club official commented. Trucks lumbered up the road at 7 mph, a string of passenger cars trailing behind. The predictable restless automobile driver swung out around the balky source of frustration, lucky if no carnage resulted. Vehicles raced around the curves, down the declines, and all too often off the edge. In the public's perception, the Ridge Route went from magnificent mountain boulevard to twisting, tortuous nightmare in a handful of years.

It had the worst safety record of any California highway. And this was at a time before there was anything like the Highway Patrol. Two Los Angeles County deputy sheriffs on motorcycles tried to keep the mayhem to a minimum, and pick up the pieces when they couldn't. In 1929 the deputies relinquished their sometimes gruesome duties to the new Motorized State Police.

The Ridge Route, designed for low volume and low speeds, soon became the most heavily used highway in the west. The road required constant maintenance and a series of minor upgrades. These were always a pace behind the spiraling increase in numbers of cars (and increasing impatience of their drivers) on the road. Highway workers continued to be housed in the Liebre Road Camp on a full time basis until the Ridge Route was replaced by the Ridge Route Alternate in 1933.

Paving, Daylighting, Desertion

It had taken one year of hard labor to do the grading and five months for the oiling of the Ridge Route. As soon as the highway opened in 1915, the motorists took to it in droves. Trucks delivered newspapers, films, and oil company supplies to the southern San Joaquin and returned with loads of milk and produce. Families took car trips that were previously thought to be too difficult.

The roadbed had plenty of time to settle in before the paving could be done, when materials and workers were again available after World War I ended. The highway was closed to

In an effort to modernize the inadequate Ridge Route during the 1920s, dirt banks were cut out, or "daylighted" and the road straightened to some degree. An entirely new road was actually what was needed.(CalTrans)

the public in February of 1919. One can only imagine the complaints that resulted from that! Scarce water and the other supplies needed for the paving job were hauled in from Lancaster in the desert. Workers were still difficult to recruit.

Nevertheless, the Ridge Route received a generous "full width" of 20 feet of cement on the 24 foot roadbed, much of it reinforced with steel, at a time when most highways were only 15 feet wide. At the same time, culverts, wooden guardrails and concrete curbing (where guardrails could not be anchored) were installed. The road reopened in November, more popular than ever.

By the time the Ridge Route Alternate and the new Grapevine Grade alignment were being built in the 1930s, steam shovels were common and mules were on their way out.(CalTrans)

In 1920, 776 vehicles per day used the Ridge Route. By 1925, the number had increased to 2,280. That would be one and a half cars per minute, all day and all night. The situation was nearly hopeless. But there could be no retreat. By this time easy and constant vehicular movement up and down the state was taken for granted.

The state gasoline tax initiated in 1923 brought a small ray of hope for the road. This new source of revenue was used for more extensive improvements, especially what was called "daylighting." In this, the banks alongside the sharp curves were cut way back to improve visibility.

A twenty person crew and one power shovel did the work. The worst of the kinks were straightened out some, or at least lost a few degrees of curvature. One of the more major improvements was dubbed the Callahan Line Change, where seven sharp turns were carved into one long 1500 foot curve. A few short passing and emergency parking areas were carved out. All just stop gap measures, to be sure.

Even more generous tax laws were passed in 1926. With secure funding and modernized equipment that greatly cut excavation costs, the highway engineers abandoned all attempts to bring the Ridge Route up to acceptable standards. They looked off toward the sunset, came down off the ridges, and drew up plans for the Ridge Route Alternate just as fast as they could.

A Highway to Remember

The survey parties set out in 1929. On the highway they staked out, the Ridge Route "Alternate" (for this was expected to be only an alternate route, not necessarily the first choice of just about everyone) average speed was expected to be increased by 50%. This "new" Ridge Route, in use from 1933 to 1971, is the Ridge Route that most of us remember:

Three lanes, then later four lanes, of overheating cars in the summer. Possibly slick surfaces, poor visibility, maybe a thin layer of snow in the winter. Trucks grinding agonizingly slow up the Grapevine. Stopping for a break, a brief romp under the large oaks at Ft. Tejon. Shooting through the Piru Gorge, unaware that this, the prettiest part of the entire drive, would someday lie deep in a watery tomb.

Another slow climb up the Three Mile Grade. Over Whitaker Summit, then trucks bearing down mercilessly upon descending the Five Mile Grade. Eyeballing those amazing truck escape ramps that could set one's imagination on edge, even if you'd never seen one in use. Down to Castaic Junction, and at last safely over the Ridge Route. We were now in Southern California proper...

The day was Sunday, October 29, 1933. The pristine 30 foot wide length of concrete shone invitingly from behind the road barriers at Gorman and at Castaic. Cars were lined up, and at 10 am the barriers were removed. From the north and from the south, the caravans ceremoniously converged at the "Channel Change" in Piru Canyon. The bands played. The speeches were made. The ribbon was cut. Another Ridge Route was opened, sending the first into obscurity.

The highway builders had gone back down into the canyons to the west and commenced work on the Ridge Route Alternate in 1927. This was the be the "final" alignment. From here on out the road could simply be widened as the need arose, or

The Ridge Route Alternate was long overdue by the time the ribbon was cut at the "Channel Change" along Piru Creek on Oct. 29, 1933. Traffic quickly plied the scenic Piru Gorge and the 1915 Ridge Route was instantly forgotten.(CalTrans)

so they said. Apparently the long-proposed damming of Piru Creek, one of the reasons this alignment was rejected in 1912, was forgotten.

The new road diverged from the old (going north) at Castaic. This time, steam shovels outnumbered mule teams. The road bed was leveled then watered, rolled and paved—no need to wait for the ground to settle. The crews proceeded from the south through Marple and Violin Canyons. They blasted their way through Piru Gorge, in the process creating the huge pyramid shaped rock that became a familiar roadside landmark. Four bridges were built over Piru Creek and a couple of others avoided by forcing the creek into a new concrete channel, the "Channel Change" where the opening festivities were held. Ten miles and 90 of those 110 complete circles were lopped off from the original route. The minimum radius of curvature increased from 70 feet to 1000 feet.

A bus rolls down the Grapevine Grade , by now (1947) a four-lane expressway. The innovative concrete barriers had been installed only the year before.(CalTrans)

The new and the old roads

converged at Gorman. It was still another year until the Grapevine section was straightened out, finally eliminating Deadman's Curve and many others. In that area complications arose by having to share the narrow canyon with the pipelines and pumping stations of two major oil companies.

The new Grapevine grade cut relatively straight down the west side of Grapevine canyon but continued to see its share of brake failures. The unlucky little town of Grapevine at the bottom of the grade was the site of a couple of disastrous truck wrecks that wiped out parts of the town. Metal barriers on wooden posts had been installed between the traffic lanes on the grade. They didn't last long with the big rigs using them for additional braking, and in the mid-forties concrete barriers (one of the first places these were tried) were installed with greater success.

By 1936 there was a 30 foot wide highway stretching all the way through the Tehachapis from Los Angeles to Bakersfield. The Ridge Route Alternate was a vast improvement but was not without its problems. It was still steep. Trucks still went out of control. It was the Highway Patrolmen's duty to cut in front of the runaways

RIDGE ROUTE STATISTICS

	1915 Road	1933 Road
Length	36.45 mi	26.85 mi
Total curvature	35,141°	2,492°
Min curve radius	70 ft	1000 ft
Maximum grade	6%	6%
Roadbed width	21-24 ft	38 ft
Total rise	4630 ft	3450 ft

GRAPEVINE GRADE

	1915 Road	1934 Road
Length	6.04 mi	5.22 mi
Total curvature	3396°	459°
Min curve radius	80 ft	1000 ft
Maximum grade	6.3%	6%

INTERSTATE 5

Maximum grade	4.5%
Max curve radius	3000 ft

65

to warn oncoming cars with lights and sirens. The one middle passing lane used by both directions of traffic caused the predictable head-ons. And another war postponed the much-needed widening.

In 1948 work began on converting the Ridge Route Alternate into an "expressway" that was complete in 1952. As much as possible, the 1933 highway was incorporated into the expressway design. A fourth lane was added, nine feet of shoulder on each side paved, and the directions were separated by a six foot strip. As planned, no major realignment and comparatively little excavation had to be done.

The cars kept on coming. By 1955 plans to convert the highway into a six-lane freeway were already being made. By the time construction was started in the early sixties it had been upped to eight lanes and a width of about 200 feet. Thus was born Interstate-5, burying Highway 99 in it's wake. In general, the Highway 99 expressway was made a part of the freeway or a parallel frontage road. In many places the four-lane expressway became one half of the eight-lane freeway. By the late sixties the last of the 99 shields were taken down.

For this latest (last?) version of the Ridge Route, I-5, it was not just a simple widening. A tremendous amount of material was moved. *27 percent more than the figure of 54.2 million cubic yards moved to construct the Aswan Dam*, according to *California Highways* (a magazine that reveled in this sort of fact.)

Some other innovations were made. Tejon Summit was actually lowered in elevation by 39 feet to reduce the grade and avoid building a costly 9,600 foot long tunnel. Along the Five Mile Grade just out of Castaic traffic lanes crossed each other and reversed at the "English Switch," southbound lanes to the east, northbound lanes to the west, to better conquer the slope. The two directions were also widely separated to prevent head-on collisions on this steep grade, as they were on the Grapevine grade to the north. Grades on the newest Ridge Route are no steeper than 4 1/2 percent.

Ten miles of road had to be abandoned altogether because it seems we did need a dam on Piru Creek. Pyramid Lake filled with water "borrowed" from northern California in 1970 and the picture postcard stretch of the Ridge Route on Highway 99 known as the Piru Gorge was forced into hiding.

Alive and Well

A surprising number of Highway 99 roadside businesses have withstood the test of time. Some of these are, The Fruit Basket in Madera ("Since 1945"), the Ramirez gas station also in Madera, the Pope Tire Company in Fresno ("Since 1918", even before all tires held air!) Savor a French Dipped Sandwich at Philippe's in Los Angeles ("Since1918"), try one of the "best date shakes in the world" at the Indio Date Shop ("Since 1932").

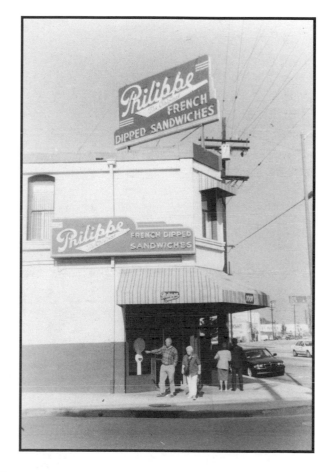

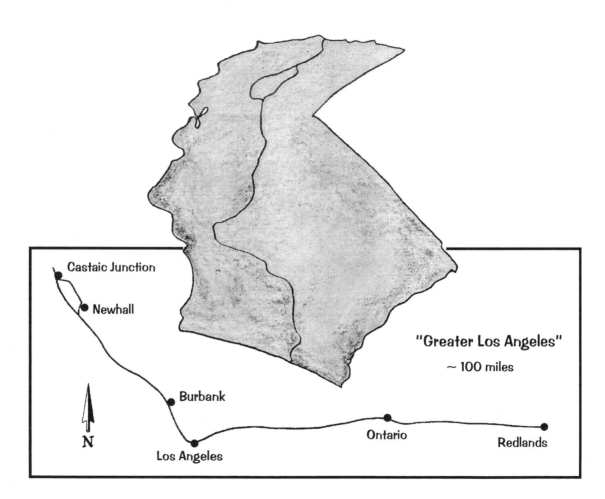

Castaic Junction

Newhall

Burbank

N

Los Angeles

Ontario

Redlands

"Greater Los Angeles"

~ 100 miles

72

CHAPTER 4

THE CITY AND THE CITRUS BELT

It's deceiving. You come down off the Ridge Route, that mountainous leg of the journey south down Highway 99 safely completed, and it seems as if you are almost "there." In reality there are many miles to go before reaching Los Angeles. Even farther to reach the "Inland Empire" in the citrus belt east of the city. And we still have over 250 miles to travel to the end of what was Highway 99 in Calexico.

Castaic was the official southern terminus of the Ridge Route. A couple of miles further down the road where Highway 126 forks off toward the coast, Castaic Junction was an anticipated post-war rest stop on the frontier of southern California; restaurants, gas stations, a profusion of shade trees and plenty of parking. James Dean ate his last meal here at Tip's Restaurant in 1955. Then he hopped back into his Porsche, turned north onto Highway 99, got a speeding ticket south of Bakersfield, turned onto Highway 466 (now called 46) at Famoso, and shortly thereafter became a legend.

Water on a Rampage

Pre-freeway Highway 99, reincarnated as the busy commuter-clogged The Old Road, is more or less intact in the few miles from the now-serene Castaic Junction south to that abomination of modern culture called Magic Mountain. The old highway crosses the Santa Clara River on a bridge built in 1928. There is a story in that date. And, as is common in California history, it relates to water.

By the turn of the century Los Angeles was getting thirsty. The city needed water to grow. The Los Angeles River was simply too fickle and too feeble a stream to be relied upon. And, the San Fernando Valley needed water to develop its agricultural potential. (Or so thought the promoters who bought up the fallow but fertile parcels at rockbottom prices.)

So the City of Los Angeles, under the auspices of Chief City Engineer William Mulholland, set out to secure a water source. The clever but somewhat deceitful scheme captured water from the slopes of the eastern Sierra. The small Owens Valley farming and ranching towns, the rightful recipients of the mountains' bounty, were left high and dry.

Ethical or not, building the remarkable 238 mile long aqueduct that opened in 1913, the tunnels and ditches and dams, was considered at that time a feat second only to the building of the Panama Canal. Mulholland was viewed as a savior.

One element of the project was the storage facility behind St. Francis Dam up

San Francisquito Canyon. On the night of March 12, 1928, the dam buckled and collapsed. Water rushed down the Santa Clarita Valley, killing hundreds of people, flooding 10,000 acres of farm land, destroying miles of highway and train tracks, ten bridges. A guilt-ridden Mulholland was said to "envy the dead."

As a result, the bridges are marked "1928" on old Highway 99.

Highway 99 on its course from Castaic to the San Fernando Valley evolved in the same manner as the Ridge Route: two-lanes, to three-lanes on a new alignment, to expressway, to freeway. But being closer to the city, from which all manner of advancements seem to emanate, the renovations through here were on a slightly earlier time table.

A Slot Through the Mountain

In this area not far from Los Angeles there remains a relic from the very earliest highway (a term used quite loosely in this case) that packs a huge visual impact: the road cut through Newhall (or Fremont) Pass known as Beale's Cut.

Beale's Cut was dug by pick and shovel in the hands of Chinese laborers in 1862. The fifteen foot wide and sixty foot deep cut was initially part of the toll wagon road franchised by Edward Beale, owner of the largest land grant in California (Tejon Ranch) and former Surveyor-General of Nevada and California. Very narrow, very deep, and with a 29% grade, it made passage over San Fernando Mountain by motor vehicle possible, but just barely. This remarkable excavation is revealing of the rough conditions faced by the early hardy motorists.

The Newhall Tunnel was the 1910 replacement for brutally steep Beale's Cut. It was designed to ease the traffic flow toward the city of Los Angeles.(CalTrans)

When the State Highway system was still merely a dream, the few automobiles there were rumbled over roughly improved wagon roads. The old toll wagon road through here became a portion of the formidable and exceedingly long desert-traversing Midway Route between southern California and the San Joaquin Valley, the most direct automobile route at the turn of the century.

This southern portion of the circuitous Midway Route which went through Saugus and Newhall and then over Newhall Pass, was later encompassed by the new State Highway when the course of the 1915 Ridge Route was laid out. The Ridge Route, of course, eventually became US Highway 99. So it can be said that the earliest version of Highway 99 went through this interesting historic landmark.

However, by the time this road through Newhall Pass, the southern extention of the Ridge Route, was declared an official State Highway, Beale's Cut had already been replaced by the nearby Newhall Tunnel built by Los Angeles County in 1910. The tunnel site is now a large four-lane-wide road cut housing what is now called the Sierra Highway. (The Sierra Highway is former US Highway 6, which went all the way from Long Beach, California to Cape Cod, Massachusetts and for a time shared a portion of its routing with US Highway 99.)

In 1939, the narrow 17 foot five inch tunnel was removed, the pass widened and "daylighted". By that time, Highway 99 had been realigned through the more direct Weldon and Gavin Canyons.

Enveloped by Civilization

Heading south from Castaic and the end of the Ridge Route proper, the old highway continued to traverse mountain and canyon along its way to the city. This portion of Highway 99 between the Santa Clarita and San Fernando Valleys was considered by motorists to be the southern extension of the Ridge Route. The same goals of a more direct and safer route that saw the Castaic-to-Gorman Ridge Route replaced by the Ridge Route Alternate in 1933 were met when the Weldon Canyon Bypass was completed in 1930, bypassing the towns of Newhall and Saugus.

The Weldon Canyon realignment of Highway 99 broke a bottleneck that had long impeded the smooth flow of traffic into Los Angeles. Narrow Newhall Tunnel was avoided and a mile or so cut off of the trip.

Just like on the Ridge Route Alternate to the north, the 1930s Weldon Canyon Bypass (or Newhall Alternate) was built as a three-lane highway (two lanes with a middle passing lane), was remodeled into a four-lane divided expressway after World War II, widened to an eight-lane freeway in the late sixties.

As it turned out, through here much of the four-lane expressway was not actually incorporated into the 1960s freeway but was left to the side to be used as a parallel frontage road. A fortunate thing too, since the former Highway 99, posturing as "The Old Road," proved invaluable in its use as a long term detour following the collapse of some I-5 freeway bridges in the 1994 Northridge earthquake.

Saugus and Newhall, those paved-over outposts of suburbia, bear no evidence of their former positions along pre-1930 US Highway 99. Considering the changes that have taken place in southern California since the towns were bypassed nearly seventy years ago, that would be expected.

At the north end of the San Fernando Valley we come across another monument to Mulholland's water scheme, and this an official one, the brass plaque at "The Cascades." Here, on November 5, 1913, fifteen thousand cars were parked. The many thousand spectators trudged up the hill, lined the concrete streambed and waited (while the bands played, the speeches were made, the salutes fired) for the gates to open and spew forth the first days' 260,000,00 gallons of Owens Valley water.

Before 1930, Highway 99 made a sharp turn across the railroad tracks on the Tunnel Station Viaduct. After the more direct Weldon Bypass was built in 1930 this bridge no longer carried 99 itself but instead marked the intersection of US99 and US6. The bridge was widened in 1936.

As a child, uncomprehending as I was of the sinister tale behind it, the frothy water on its long, controlled tumble down the mountain was always an impressive sight. All a part of the Water Department's public relations plan to be sure; the water could as well have continued its journey in an underground pipe.

Former US Highway 99, now in the form of San Fernando Road, resumes its position alongside the railroad tracks all the way through San Fernando Valley. Highway and railway parted ways soon after leaving Bakersfield. The train, of course, could not "make the grade" over the mountains traversed by the Ridge Route, instead following that roundabout desert circuit.

It is a long, straight industrial approach to the City of the Angels; train tracks, warehouses, factories; the necessary workings behind the metropolis. It is difficult to imagine that this valley was once "the market basket of Los Angeles"...Along the way are found some classic motels, a few of them (not enough!) with nice old neon signs, a car wash and other roadside miscellany.

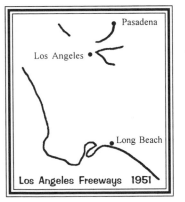

Los Angeles Freeways 1951

A Love/Hate Relationship

The historical routing of highways through urban areas is more difficult to trace than the course through less populated areas. "Put the State Highway right down Main Street," the politicians cried. Then, in succession, "Too congested! Move it over a block," and "Too dangerous! Go around us."

The consequence being that a highway's track through a city might change frequently. Then there were the "Alternate" routes (bearing the same highway number) that were sometimes designated to either avoid or lead directly to the downtown area.

Add in the fact that even after US Highways were assigned their official numbers in the mid-twenties, these numbers were not always printed on the road maps. Or, confusing matters even more, a US Highway such as 99 might be labeled with one of its State Highway numbers, which differed from the US Highway number even through it was the very same road!

Complicating matters even more, several numbered highways might converge and share the same pavement on the urban approach and regress, as well as through the city.

As for Highway 99 on its windings through the great sprawling city of Los Angeles, the records are confusing and incomplete. And in a sense, it really doesn't matter which surface street the highway was routed on in this year or that. The fact is that pre-freeway era highways always lead into the heart of the city. On the way in and on the way back out, subdued by the urbanization and by sharing the roadway with other numbered highways, it seems that they lost some of their individual identity.

After the straight shot down San Fernando Road, Highway 99 took a sharp jog to the west to cross the Los Angeles River, surely one of the most derided rivers in the world. The crossing was made on one of the graceful concrete bridges that was built early in the century. It is difficult (given the traffic conditions and lack of parking) to get enough of a view to appreciate the bridges' beauty.

A rare unobstructed view of the Los Angeles City Hall is seen along the original Highway 99 routing through "new" Chinatown.

The nine L.A. River bridges were built from the teens through the thirties as something of an ensemble, each within view of the next, complimenting in design the next one down the line. Unfortunately, some of the ornamentation from the earlier era has been removed but the integrity of the designs remain.

Highway 99 looped through downtown Los Angeles. It went through Chinatown, skirted the Civic Center, passed by the red-tiled quasi-Mexican Olvera Street in historic Pueblo de Los Angeles and the spectacular marble-floored Union Station before turning sharply east. The highway again crossed the river on another of the concrete bridges and headed toward the outlying citrus belt, the "Inland Empire."

Which particular bridges and surface streets carried 99 into and then out of downtown Los Angeles varied with the alignment of the era. At different times, Highway 99 shared segments of its path through the city with Highways 66 and 101, as well as Highways 6, 60, and 70.

Los Angeles, being the quintessential automobile-oriented city, built some of the nation's earliest freeways. These were first termed "parkways" in recognition of designs and landscaping created to beautify the urban environment. The very first was the six mile long Arroyo Seco Parkway, now the Pasadena Freeway, that opened in 1940.

Soon thereafter, a 1943 wartime project saw Highway 99 on its southern departure from downtown

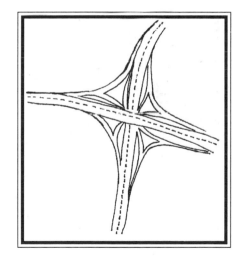

Four Level Interchange in Los Angeles, the first of its kind, 1953.

83

Los Angeles (actually east but this was a north-south highway, remember) upgraded to freeway status.

So at the beginning of its run from the edge of the Los Angeles River to the Mexican border, the portion of Highway 99 that had been carried on Ramona Boulevard, after 1943 was covered by the new Ramona Parkway. This is now the San Bernardino Freeway, I-10.

The earliest freeways were only short segments. Before long the intricate freeway system snaked all over the place and pleasing design elements were less important. The freeways were quickly extended and connected. For a time, the Highway 99 alignment circled on one of the clover leaves on the famous "four level interchange" near downtown L.A. Later, as part of the Golden State Highway, it avoided the downtown altogether. And by 1967, the US Highway 99 shields were taken down from the last Highway 99 alignment, the Golden State Freeway, Interstate-5.

From Citrus to Suburbia, The Inland Empire

With the familiar shape of the Los Angeles City Hall receding in the skyline behind us, we resume our journey down Highway 99 aided by the 1939 *California, A Guide to the Golden State.*

Monterey Park ...has many large estates along the rolling hills of its western extension, as well as rows of small simple houses and bungalows...It is in an area of rich, sandy loam that has been transformed by irrigation into a highly productive walnut, avocado, citrus fruit,

Idylic scenes of citrus and palm such as this one in present-day Redlands were once the norm where Highway 99 traversed "The Inland Empire."

berry, truck garden, and poultry district...

On US 99 east of Covina Blvd. truck farms and chicken ranches diminish as unbroken lines of orange trees and walnut trees appear on either side of the highway...West Covina...consists of large lemon, orange, grapefruit, and walnut groves, each with its rambling frame or stucco home built well back from the pepper and eucalyptus-lined avenues...Pomona...is a shipping point for 30,000 acres of citrus groves...

The city spread like a cancer in the post-war years. The highway leaving the city filled with motels, trailer courts, and eateries, many of them still there. But what dominated the windshield vista when driving down Highway 99 up through the 1950s was mile after mile, acre after acre of vineyards and citrus groves, an expanse that today it is impossible to fathom.

The glossy green quilt cloaked the land, extending up into the foothills above the San Gabriel, Pomona and San Bernardino Valleys. The occasional small remnant grove or backyard orange tree, field of untended grapevines, or row of eucalyptus that once served as a windbreak to block the strong Santa Ana winds, sadly, is about all that remains.

In this country, water tapped from the snow-capped mountains to the north was shunted into irrigation ditches and the land sold off in parcels by sometimes unscrupulous salesmen ("where are the paved streets and water system you promised?") beginning in the late 1800s. As trees and vines came into production, cooperatives were formed to care for the groves and vineyards and to market the produce. And so the lush, well-watered southern California that attracted the hoards of emigrants, which ironically led to the destruction of this paradise, took shape.

Highway 99 went through the town of Ontario on Holt Boulevard. Ontario is a good example of this planned agricultural community-making process. The successful town was founded by the Canadian (hence the name Ontario) Chaffey brothers in 1882. The irrigation system was installed as promised, the tracts were laid out and sold, the trees were planted and came into production, the citrus was sold at a profit. Ontario was, in fact, the "Model Colony."

Along the way to success, area citrus cooperatives staged massive publicity campaigns. Such tactics as distributing California oranges from specially decorated trains at mid-western whistlestop towns not only helped make orange juice an irrefutable element of the American breakfast but also set winter-chilled minds to thinking, "Why don't we just move there?"

And move they did. Initially they came to grow the oranges. But gradually, when there wasn't room for both the oranges and the people, or rather, when homesites were worth more than citrus groves, the oranges lost out.

Ontario has long held an event called the All-States Picnic under the stately pepper trees of Euclid Avenue to celebrate the diverse origins of the local settlers. But by the post-war era, more and more of the residents were homegrown. By the time I was an attendee of the picnic in the mid to late fifties, every state

had its allotted picnic tables, but the "California" area was far and away the largest.

Leaving Ontario, the next five miles of highway were once lined solidly by the Guasti vineyards, at 5,000 acres, *the largest vineyard in the world.* The quaint little Italian town that housed the winery and distillery is still there, a tranquil island amid the spreading suburbia.

Then the highway skirted Mt. Slover *with craggy rock outcroppings, jutting up sharply in a white mist of smoke and dust...The whole hill is gradually being carted away and ground up.* The cement mountain has steadily diminished in height, no doubt in part due to the early paving of state highways with cement.

And on to Colton, *an industrial and railroad center.* Then across the Santa Ana River, *a winding trickle in a broad, brush-lined bed.* Beyond which *the highway strikes through a jungle of auto courts, roadside cafes, and fruit and vegetable stands; through fields with scattered trees and ranch houses where cattle graze; past barns, fenceposts, and windmills. Orange trees reappear, clustering thick and leafy along the route,* to Redlands.

Redlands was on the outer boundary of the citrus belt. It was full of fruit packing houses, was considered something of a cultural center, and acknowledged to be an attractive town (still is). Leaving it, *US 99 winds past hedges of roses through Reservoir Canyon.*

The road begins to climb again. Highway 99 approaches the desert realm, the last segment of the long journey to the end of the road.

Then and Now

Many of the buildings that long ago housed various highway enterprises, those that haven't been mowed down by newer roadside businesses, are in use to this day. Some are obvious; others require some keen observation to find.

Along Holt Boulevard, the former US Highway 99 in Ontario, the Jiffy Lunch was typical of a restaurant called a "luncheonette". Today the same building, only slightly altered, houses a $4.95 All You Can Eat menudo restaurant (p.90-91). This is common along the old highway, and indeed throughout much of the state. The restaurants which once dispensed hardy American fare now serve Mexican food or "Asian fast food" a clear example of the changing ethnic makeup of the state's population.

An occasional small fragment of untouched very old highway can be found here and there as well. Near Galt in the north San Joaquin Valley, a 1997 photo of the circa-1910 concrete road is difficult to distinguish from the one taken eighty years earlier (p.92-93). This short piece of forgotten highway leads south to the county line at the crossing of Dry Creek. The bridge no longer stands.

(Ontario Library, Model Colony Room)

Ontario, California. Once Serving Meatloaf, Now Serving Menudo

(CalTrans)

The same piece of highway near Galt, very early and very late in the 20th century.

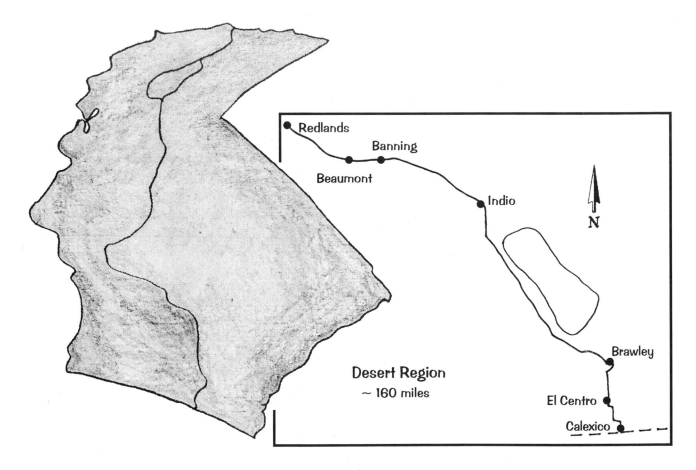

Redlands

Banning

Beaumont

Indio

Brawley

El Centro

Calexico

Desert Region
~ 160 miles

N

CHAPTER

5

DATES, SAND, AND SALTY WATER

And so Highway 99, which began far up the highway in the fog-shrouded northwest, traverses the California desert on its very last leg through the American west from Canada to Mexico. No more mountain ranges to cross, no more large cities through which to navigate.

This north-south highway diverged sharply from its aim straight to the south by following a southeastern course out of Los Angeles. That directed the road away from the bulk of the population and, after leaving the land of orange trees, across the sandy miles that would have been an absolute no man's land if it were not (again) for imported water.

The Last Mountain Pass

The two highest mountains in southern California, San Gorgonio and San Jacinto, loom above the highway, a mountain on either side, as it proceeds out of the orange grove havens and over the San Gorgonio Pass, the last mountain pass on Highway 99.

On the northern ascent of the pass the twin towns of Beaumont and Banning (*site of several sanitoriums*) were known for their dry, healthful climate and surrounding deciduous orchards. In Beaumont *the blossoming of the trees is celebrated with the Beaumont Japanese Cherry Festival suggested by the cherry blossom fetes in Japan. Long lines of motor cars drive along the blooming orchards of cherries, almonds, apples, peaches, and plums.*

Cabazon is *sandblasted and weatherworn by the almost continuous winds through the San Gorgonio Pass...Most of the population is employed by the railroad.* The wind

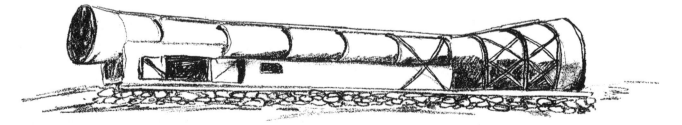

The Oliver Electric Power Generator, 1926

96

whipped pass is topped soon after leaving Cabazon. Local Indian legend has it that the day the wind stops blowing through San Gorgonio Pass is the day the world will come to an end.

At the pass and eastward, unevenly spaced lines of wind generating machines dot the surrounding hills to capture the free energy. The idea of nabbing that obvious energy source is not a recent one. A large curious contraption erected beside the old highway in 1926 once caused motorists to slow down and have a look, maybe even pull out their checkbooks. The machine was called Oliver's Electric Power Generator.

The purported plan was to build many more Oliver Power Generators and sell the electricity once some investors could be found. Now this ten ton funnel-shaped machine did in fact generate electricity. But it was a paltry fifteen kilowatts worth, not enough to sell to the nearby towns which in any case already had their own generating systems in place.

Putting up the odd metal device alongside Highway 99 was a way to attract investors in "Oliver Power." At least that part of the plan worked. It seems that indeed may have been the only plan. Dew Oliver, promoter and real estate salesman, eventually spent time in jail for fraud for this scheme. The prototype Oliver Power Generator remained in place, a roadside curiosity for many years, until it was scrapped in 1942.

Coming down off the pass the highway crosses the surprising Whitewater River on a long eight-spanned concrete bridge. *To this day even one does not know when the dry stream bed may change into a torrent and monopolize man's puny roadway that crosses it so apologetically.*

The Date Capital

What was Highway 99 descended into the Coachella Valley. The land gets lower and sandier. In the town of Indio *the highway winds through the town between rows of service stations, garages, cafes, and other appurtenances of the tourist trade.* Indio was once *one of the lustiest and toughest of desert settlements* but is now a mild-mannered Date Capital of the Nation. Not to be overshadowed, the town of Coachella just down the road is the Grapefruit Capital of the World.

It seems that nowadays just about any imaginable food item is available for purchase at a good grocery store. That what was once exotic is now commonplace. There was a time when produce such as dates were thrilling and difficult to obtain. It was worth a drive to the desert to pick some up, and a rare treat to send in a Christmas parcel.

The roadside was once replete with date shops doing a brisk business. When business slowed during WWII austerity, some of the slack was taken up by the many servicemen stationed at nearby bases. They just loved the date shakes.

A handful of the roadside date shops are still in business. Date palms were introduced here in 1912 by the Department of Agriculture. The government officials

An undated photo shows Highway 99 near its end in Imperial County. (CalTrans)

Classic Date Shake

Date Mix: 1/2 pound chopped dates
           1/4 cup water

Let soak overnight. Stir up with a fork in the morning.

For one 16 oz shake:
     1 scoop (~1/4 cup) date mix
     1/2 cup milk
     4-6 scoops vanilla ice cream

Mix in a blender, and enjoy!

Up-Dated Shake

1/3 cup dates
3/4-1 cup 2% milk
1/2 pint vanilla frozen yogurt
dash of nutmeg and/or cinnamon

Mix in a blender, and enjoy!

---

had collected and experimented with many date varieties from Arab countries in an effort to put our "wasted" desert lands to work. This seems to have been a success; King Feisal of Iraq, on receiving a gift of Coachella Valley dates, is purported to have said, "We, who have been growing dates for centuries, have never seen such fine dates in our own country."

The Accidental Sea

Imagine traveling old Highway 99 across this flat expanse south toward the Imperial Valley and the end of the highway at Calexico, temperature 110 degrees (or more!) and air conditioning not yet invented. The canvas bag would be hung, dripping, over the radiator and the windows would be wide open, the wind creating an illusion of coolness. It would be a barren journey indeed if not for the shimmering water of the Salton Sea off to the east. An unusual lake it is, too; little vegetation around it, very salty water, its presence at this time an accident of man and nature.

The Imperial Valley is a part of the ancient Colorado River delta and as such is very fertile, though dry. Ambitious turn-of-the-century land developers recognized this. They were tempted by the logic of diverting water from the Colorado River only sixty miles to the east into a canal. The Imperial Valley could bloom and prosper.

No permission was granted to either take the large quantity of river water nor to dig the canal through a foreign country (for a portion of the canal was in Mexico.) These questions were not even raised until later on when the competition for water intensified.

Nevertheless, the first canal was easily dug in 1900. By 1905, 67,000 acres were under cultivation and the population had jumped from near nothing to 12,000. That winter a flooding Colorado River (not yet controlled by Hoover Dam) broke through the canal and tried to resume a course that had been abandoned four hundred years earlier. The Salton Sink filled with water, thousands of acres were innundated or eroded away, railroads and highway were washed out.

It took two years to stem that disrupting flow. Eventually, the canal system was revamped and the farmers recovered, the better to provide us with our winter salads. The All-American Canal (entirely on U.S. soil) was proudly and legally built. The Salton Sea was cut off from its Colorado River source but continues to be fed by irrigation runoff.

Below Sea Level

Shortly after leaving Indio, Highway 99 dropped below sea level and stayed that way until its termination point at Calexico, which sits at a grand elevation of five feet above sea level. Barring the heat, highway construction must have gone swiftly along this portion. The ground was flat and dry, no bridges to build except for a few over narrow canals. Neither was it plagued with deep sand such as there was on the highway to Yuma, Arizona that required building a wooden plank road to sit atop the shifting sands.

The series of agricultural, below-sea-level towns were then (as now) populated by many Mexicans and Mexican-Americans who worked the fields, *following the ripening crops in a great, nomadic trek, living in makeshift huts and "jungles."* This was a center of unrest among agricultural workers long before the days of Caesar Chavez. Following World War I, Hispanics were joined by *Filipinos, Hindus, Japanese, and Negroes* and later, white dust bowl refugees. The toxic brew of poor living conditions, low pay, no sanitary facilities, and less complacent workers led to a strike in 1934. One day, 8,000 lettuce pickers walked off the job. Things got ugly, then bloody. The government mediators that were called in eventually gained meager concessions for the workers. This problem is an ongoing challenge in our state.

Brawley had *a plaza filled with tropical trees, flowers, and plants.* Then through fields of cantaloupes, lettuce, alfalfa. Near Imperial *a common sight is signs announcing "Cotton Pickers Wanted."* El Centro is a little city *with shaded patios and carefully tended flower gardens...Here as in cities much nearer the equator the stores have arcaded fronts over the sidewalks and many of the houses have air chambers under their roofs.* It was also at the crossroads of Highway 80 (supplanted by I-8) which headed west to San Diego or east to Arizona.

This final stretch of Highway 99 was even more of an agricultural/industrial road than it was coming through the Central Valley. Up there, the produce trucks and agricultural equipment on the road were accompanied by motorists going to or from Los Angeles, San Francisco, Sacramento. Down here, there were no major destinations beyond the Palm Springs turnoff near Indio. Wide-open Mexicali across the border from Calexico was a draw during Prohibition, but most Californians in search of a good time went to more convenient Tijuana.

At the International border of Mexico and the United States of America, in the town of Calexico, California, US Highway 99 completed (began) its 1,500 mile north-to-south cross-country journey. The number "99" has been forgotten by many.

For those in the know, the still visible remnants of Highway 99 represent the decades of this century when everything seemed possible, when growth was good, when the future was limitless. As things happened, the history of this highway and the country it went through has not always been irreproachable. But it's where we've been.

Roadside Attractions

Being the utilitarian highway that it was, Highway 99 probably had fewer than its fair share of "roadside attractions." But a roadside attraction, something exhibiting enough appeal as viewed through the front windshield of a moving car to incite the driver to pull over, could be as unassuming as a fruit stand with a shady vine-covered picnic area and a couple of old farm implements on display. One step up and away from that were the many "Giant Orange" restaurants which lined Highway 99. Those large, garrish cement-over-chicken wire oranges that dispensed fresh orange juice and hot dogs were a huge draw. That in-your-face style of "programmatic architecture" common in the 1930s and 1940s, was born in response to automobiles. A roadside business has only a few brief moments in which to entice a motorist to stop. An outlandish building in the shape of an animal, piece of fruit, a shoe, or any other object is one sure way to be noticed.

Pollardville near Stockton

A couple of more typical Roadside Attractions still prosper along Highway 99. The one known as Pollardville Ghost Town evolved from a simple chicken restaurant. A la Knott's Berry Farm, but seemingly still in 1952.

Pollardville is located between Lodi and Stockton on the northbound Highway 99 frontage road. The old two-lane 99 alignment through here is covered by the current 99 freeway's southbound lanes. Interestingly, Pollardville does not have an adjacent offramp, a usual cause of roadside business failures.

Neil Pollard tells the story of his parent's chicken ranch in Castro Valley in the 1930s. To add to the family income, his mother started selling eggs and chicken sandwiches. In 1944 the family opened a restaurant in Stockton. Then in 1947 they decided to move out to the highway and start a chicken restaurant. Everyone thought they would fail "way out there in the boonies." That early restaurant was built on leased land on the west side of the highway.

They did not fail and soon were able to buy the land on the other side of the highway. So they pulled the building across the road and they've been adding on to it ever since. The first "attraction" other than a hardy chicken dinner came with the purchase of a collection of "old stuff" to add to the ambiance of the restaurant.

Now there is a wonderful collection of farm and mining equipment on display around the small mining town out back. The town itself is made up of authentic buildings moved from Sierra mining towns. There is a narrow gauge railroad, a landlocked steam paddleboat, and a steady stream of birthday parties and other festivities taking place in Pollardville. It's a favorite stop for Canadian snowbirds on their way south in the fall, and again on their way north.

If the sharp peaks on the roof of the Pollardville restaurant seem odd, there is a story behind that, too. The restaurant building burned in 1984. An empty Polynesian restaurant was moved down from Stockton and redecorated. It seems that Pollardville is a survivor.

Further down Highway 99 in a section of north Fresno known as Highway City we find the Forestiere Underground Gardens. This site, on Shaw Avenue just off Golden State Avenue (the old 99) is now on the National Register of Historic Places. It began not as a roadside attraction but as one man's inspired dream.

The kitchen was cool even in the middle of a sweltering Valley summer.

Baldasare Forestiere was the creative genius who dug his way down and out of the Central Valley heat. The Sicilian immigrant started digging his underground domain in 1906. With an energy and creativity unfettered by family responsibilities (he was unmarried) he worked on his subterranean escape for forty years.

Forestiere dug up the surface hardpan in rough blocks and used it for building material down below. He planted a huge variety of fruit trees in the softer, moister depths, cutting circular openings to the surface (which he covered with sheets of glass in the rainy season) for light. When he died in 1946, he left his five-acre multi-level labyrinth of living chambers and gardens, arched doorways, ponds, and an 800 foot auto tunnel to the care of his brother Giuseppe.

Giuseppe and his offspring and their offspring took the responsibility bequeathed them to heart. The family has sacrificed a lot of time and money in keeping the Underground Gardens alive. The Gardens were opened to the public after the creator's death and quickly became a popular roadside attraction. Their survival to this day seems a miracle, given the obvious value of the land for other purposes. The Forestiere Underground Gardens are a small island in the midst of a fast food jungle.

Baldasare grew a variety of fruit trees in his underground gardens.

The first routing of Highway 99 went right through downtown Los Angeles and right by Olvera Street in the old Pueblo de Los Angeles. It was a natural place to stop, a draw for tourists then and now.

This is where Los Angeles began and is one of the oldest streets in the city. In 1930 one block of the street was revitalized into a Mexican marketplace, replete with sidewalk shops selling handicrafts, Mexican restaurants, and historic adobe buildings. It might be argued that currently other parts of LA are more authentically Mexican, and they come by it quite naturally (given the large Hispanic population.) That Olvera Street represents a romanticized version of old Mexico and an old Los Angeles that never was. But this is a genuinely historic place, and the souvenirs available on Olvera Steet are no more expensive than they are in a border town in Mexico itself.

An aptosaurus near Cabezon seems to be browsing on the desert brush.

The two larger than life-sized dinosaurs at the Wheel Inn truck stop in Cabezon are not Highway 99 roadside attractions per se. The steel and concrete creatures were not constructed until the early 70s, well into the freeway era. We give them a nod because the specter of these roadside monsters looming ominously by the Interstate brands them as the sort of thing that would have been the ideal roadside attraction of the two-lane highway era. At the same time they are authentic representations of real prehistoric species.

A museum and gift shop reside in the belly of the aptosaurus. The tyrannosaurus was intended to have a slide running down its tail but it was never completed. Creator Claude Bell died with his dream incomplete, a wooly mammoth and a host of other creatures still on his drawing board.

Stop! Tour Ahead!

(1932, CalTrans)

PART

2

A TOUR THROUGH CENTRAL AND SOUTHERN CALIFORNIA

State Capital

Much has changed in Sacramento since its beginnings as the major supply center for Mother Lode miners during the 1849 Gold Rush. As the most important city in the state, it was chosen to be the State Capital, and construction of the lofty capitol building began in 1860.(p.112)

Parks and government buildings and museums are integral to the capital, but Sacramento is located on an inland seaport at the confluence of the Sacramento and American Rivers and has always had its industrial side. (p.113)

The changing makeup of California's population is illustrated by this scene: Hindi movies are shown in a classically styled old movie theater that is shaded by a Spanish language AIDS education billboard along Stockton Boulevard, old Highway 99. (p.114)

Galt

Near the town of Galt, the Highway Department made it quite clear where one foreman's responsibilities ended and another's began. The highway through here was designated as Route 4 by the California State Legislature in the teens. It acquired a second number, 99, when becoming a U.S. Highway in 1926. (CalTrans p.115).

Lodi

In the early 1900's many towns built arches to celebrate an event or to welcome visitors. The earliest arches often faced the train tracks that first brought the travelers, while later arches spanned the highways. Lodi's lath and plaster arch is the oldest standing welcome arch in the United States. It was built in 1907 as part of a grape harvest festival.(p.116)

Stockton

Stockton is now a large city crisscrossed by freeways, but some of the businesses along the old highway have survived. This gem of an old sign exhibits well-earned wear and tear.

Manteca

Progress along Highway 99 was slowed by all of the activity at a San Joaquin County "borrow site" in the early fifties. (CalTrans)

A beautiful brick building fronts old Highway 99 in the town of Manteca, an important crossroads. From this town, a turn onto Highway 120 took motorists east to Yosemite and the High Sierra.

Salida

    This small typical valley town, like so many others, had its roots in agriculture. Yet the highway traffic appears to have been aptly served with gas stations, a tourist court, and a luncheonette (CalTrans).

Modesto

The Seventh Street Bridge, or Lion Bridge, crossed the Tuolomne River at Modesto's southern entrance along what was the first State Highway routing through this city. It was built in 1916 and is a fine example of the decorative concrete bridge design that flowered in steel-starved but cement-rich California.(p.121)

Modesto jumped on the arch bandwagon in 1912 and decorated theirs with a slogan that aimed to illuminate (both metaphorically and with electric lights) the area's benefits. Often referred to as the Prosperity Arch, it is the oldest standing slogan arch in the country.(p.122}

Livingston, Atwater

Older Mom-and-Pop motels advertise all of their amenities in a valiant effort to attract customers. The Town and Country sits in a once prime location near what was for years the very last stoplight on Highway 99, in the town of Livingston. The town was just recently bypassed by a long-awaited segment of freeway that eliminated the dangerous, often foggy intersection. It's likely that the "no vacancy" sign will seldom be lit from now on.

A motel in Atwater took the notable highway number for its name.

Merced County

In the early teens the State Highway (not dubbed "99" until 1926) was paved in narrow single slab of concrete. By the time the 65th Cavalry moved through on Highway 99 in April, 1941 the highway had evolved to three slabs of width. (CalTrans)

Chowchilla

Flooding is a frequent occurrence in the Central Valley. One of the largest floods happened at Christmastime, 1955 but it appears that a few people still braved the highway. Here in Chowchilla, Highway 99 follows the railroad tracks, as it does along much of its route. (CalTrans)

Orange Orbs

The sad remains of a once thriving Mammoth Orange have surely seen better days. This one served the southbound side of a busy four-lane expressway portion of Highway 99 just south of Chowchilla. (p.126)

Another on the east side still does a thriving business, although the building has been remodeled so that the big orange is less obvious. The restaurant is a happy reminder of the time when a stop at one of these distinctive orange balls was eagerly awaited by the back seat occupants. The first Orange was built in the mid-20s in Tracy, CA, the beginning of an early restaurant chain that was widely imitated throughout the state. Orange juice based drinks, hot dogs and hamburgers were the standard fare.(p.127)

126

127

Madera

Hardy early motorists were used to delays. This 1914 Madera County driver was probably patient with the construction ahead indicated by the sign, for soon the highway would be free of dust, mud, and chuckholes. The State Highway System had been mapped out since the turn of the century but no money was available for actual construction until after the first Bond Act of 1909. Groundbreaking took place in 1912 and the following three years brought a flurry of activity. (CalTrans p.128)

An olive-loving kid reigns over a deep green olive orchard. He appears to have been transported from the 1950s.(p.129)

The Pine and the Palm

Just below Madera, a pine (representing the north) and a palm (representing the south) were planted beside the road years ago to mark the half-way point of Highway 99 through the state.

Rumors that this also marks the geographic center of the state are off by about fifty miles. CalTrans heeded the nostalgic outcry of history buffs in 1990 and changed their plans for cutting down the trees in the process of erecting a traffic barrier.

Fresno

Travelers in the 1940s-1960s heyday of Highway 99 had myriad choices of motel accommodations and eateries. They were aided in decision making by the lure of distinctive neon signs and blinking arrows. Some establishments have fallen into disrepair and abandonment; others hang on, still sporting classic original signs such as Fresno Motel's bathing-capped diver (even with the swimming pool long ago filled in).(p.132)

132

The actual faces of Manny, Moe, and Jack are a rare sight anymore, but in Fresno they beam out over what was once the highway route through the town.

The railroad and Highway 99 followed much the same course and occasionally crossed over each other, a situation that raised concerns of traffic safety. As our highways became more sophisticated, "separation structures" were built to keep the two modes of transportation apart. In this structure on the north side of Fresno the train tracks take precedence; the highway and a pedestrian walkway go under the tracks through a subway built in 1932.(p.134)

Fowler

The small hamlet intended not to be overlooked in this long valley that is chock full of somewhat similar small agricultural towns by erecting a lighted sign on Highway 99.

Selma

The "no" in its "vacancy" is long gone from the Selma Motel sign, but many older Highway 99 motels still hang onto life by a thread. Some have been turned into apartments for the increasingly transient population.

Water Towers

Without water, there would be no agriculture and no agricultural communities in the Central Valley. The Wright Act of 1887 established the legality of forming irrigation districts. This in turn led to the complex systems of dams and canals that turned this valley into one of the world's most important ag centers. Being on the flats, the Valley towns of necessity store their water supply in high towers that help define the towns' character. Those in Galt(p.137) and Manteca(p.138) are utilitarian, Lodi's is in a Greek style(p.138), Kingsburg's is in the form of a huge decoratively painted tea pot that capitalizes on its Swedish Village theme(p.139). The Old Fresno Water Tower (no longer in use) is a beautiful brick structure dating from 1894 that is on the National Register of Historic Places(p.139).

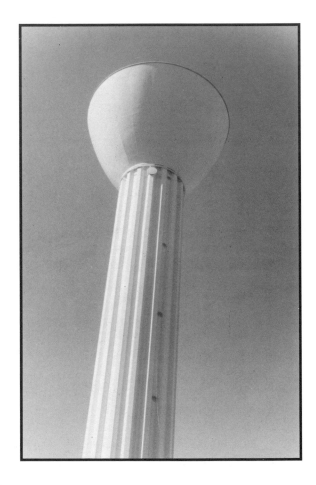

138

139

More Roadside Attractions

If the need for a quick palm reading arises when traveling along old Highway 99 below Fresno, Madam Sophia will take care of it. Perhaps most of the potential customers have taken to calling psychic hotlines, but a few roadside palm readers are still found.

A more recent 99 attraction is the airplane display at Mefford Field near Tulare. It is visible from the freeway and located at a convenient off ramp for leg-stretching.

Take Your Time!

It was a long trip down the Valley to Los Angeles at only 40 MPH, but there was a solidarity at the home front behind the WWII war effort that is hard to fathom today. (CalTrans)

Tulare

The milkmaid statue in front of Adohr Farms illustrates one of the major agricultural enterprises in this region. Rumour has it that the company name is in fact the name of the founder's wife spelled backwards.

Abandoned independently owned gas stations such as this one in Tulare are common along the old highway.

Delano

The flooded highway near Delano slowed but did not stop the motorists in March of 1952. (CalTrans)

Bakersfield

...was the last major stop before the Grapevine Grade and was full of motels to meet the travelers' needs. An arched walkway bearing the town's name crosses busy Union Avenue (old Highway 99). It was erected in 1949 and once connected the two sides of the popular Bakersfield Inn. With 400 rooms and 414 palm trees, it was both a favorite local gathering place and famous tourist stop.

Valley's End

The humble remains of a small auto court at the foot of the Grapevine Grade once provided a welcome respite for the winter traveler. Even this far south, the mountains separating the Central Valley from Southern California can at times be nasty traveling.

Neon Arrows and Stars

Before leaving the Valley, let's take a look at some of the interesting neon motel signs that have flickered by the windshield...

There once was a time when each mom-and-pop motel and cafe had its own distinctive neon sign designed to lure the travelers. Later, and still, those signs that are now considered by many to be an art form are discarded with reckless abandon in the interests of "modernization."

From the 1940s through the 1960s the art form flourished. Early renditions sung the praises of "steam heat" and "flush toilets." Later it was "TV" and "pool." The wonderful array of colors and motifs brought a variety to the roadside environment that is missing today. It's the same Denny's and Motel 6 sign in Sacramento as it is in Fresno.

One can imagine the tough decision it must have been, to order up that gas-filled glass-tubed sign that was to serve as the main advertisement for the forseeable future. It seems there were many common themes.

The western motif (cowboys with lassos shaping the name of the motel, Mexicans snoozing under a tall cactus), the palm tree motif (even well out of the range of living palm trees), the beach motif (The Sands, The Capri, The Driftwood, but the ocean many miles away), and my personal favorite, the Sputnik-era space motif.

149

There was also a liberal use of arrows: fat, skinny, long, short. The arrows with their bright rippling lights directed the confused after-dark driver right into the parking area or to the motel office.

By the 1960s chain motels had become a threat to the smaller establishments. Most who could afford to took the old signs to the dump and erected new, nondescript, understated signs. And the trend continues.

Fortunately for us, not everyone replaced their signs. Many of the old motels that served the traveling public so well and have sunk to some level of seediness, by accident or design, still retain their artful signs.

The Ridge Route

A now-quiet segment of the 1915 highway known as Deadman's Curve is isolated to the side of Interstate 5. This was technically a part of the Grapevine Grade, but the entire mountain route was commonly called the Ridge Route, as it still is to this day.

The Granite Gate looks down off the ridge toward where the replacement highway, the Ridge Route Alternate, was built in 1933, and later Interstate 5. Motorists could only travel this road at 10 to 15 mph. Anticipated landmarks such as this marked their slow progress and were even featured on postcards.

All that remains of the once lively Tumble Inn is its foundation. By the mid-1930s, almost all of the traffic had shifted to the Ridge Route Alternate and establishments such as this were quickly abandoned.

A desperate attempt was made to improve the speed and safety of the Ridge Route in the 20s. Concrete curves were widened out and then filled in with asphalt.

Ridge Route Alternate

The Alternate between Gorman and Castaic was built down off the ridges miles away from the original 1915 Ridge Route and traversed scenic Piru Gorge. Most of the pretty canyon and all but one of the old highway bridges was submerged by Pyramid Lake in 1970. Several miles of old highway below the dam are still accessible.

(CalTrans)

Four in One

A forgotten slab of concrete near Tejon Pass sits above its descendents: Interstate 5 and the post WWII four lane expressway. This double slab piece was poured in 1923 directly on top of the original 1919 single slab. The 1930s "Alternate" was obliterated by the 1951 expressway.

Castaic Creek

Finally over the Ridge Route, the old highway went through the Santa Clarita Valley. The remains of the original 1916 bridge are visible in the Castaic Creek streambed. The wreckage is backed by a replacement bridge now crowded with commuters. Just down the road, the bridge over the Santa Clara River was washed out by the San Francisquito Dam disaster in 1928.

Beale's Cut

It's a short walk to this amazing piece of the first wagon road over Newhall Pass. With automobiles becoming more common, this short but steep byway was replaced by nearby Newhall Tunnel in 1910. Highway 99 followed this route until the 1930 realignment through Weldon Canyon called the Newhall Alternate, or Weldon Canyon Bypass.

The Old and the New

Several modern freeway bridges straddle sedate old Highway 99 just north of the San Fernando Valley near its junction with the Sierra Highway. The old highway served as an important detour after a 1994 earthquake proved the new structures to be more fragile than they appear.

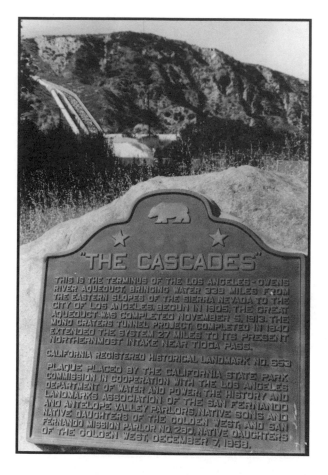

"The Cascades"

The refreshing view of frothy Owens Valley water tumbling freely down the hillside nearing journeys-end seems to announce to highway travelers their arrival in thirsty Southern California.

San Fernando Valley

...brims with motels that survive despite through traffic now taking the interstates.

Today's taco stand first saw the light of day decades ago in the form of a roadside luncheonette in the San Fernando Valley...

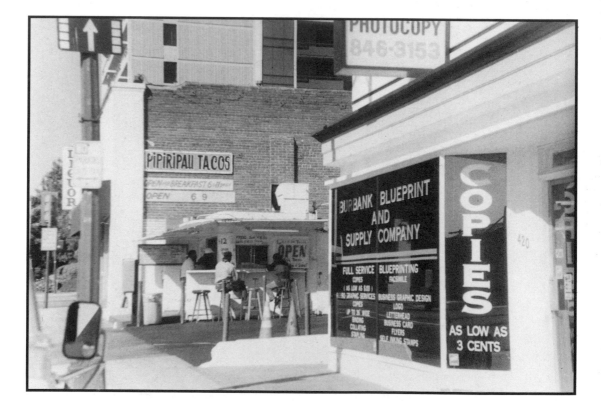

Los Angeles

The original Highway 99 routing went right into the heart of downtown Los Angeles and passed by Union Station. The architectural masterpiece with marble floors, arched windows, and a 50 foot ceiling was a bustle of activity in the 1940s and 1950s. It was erected in 1939 on top of Old Chinatown.(p.163)

It's a shame that the original 1910 ornamentation was removed from the Main Street Bridge crossing the Los Angeles River. The nine concrete L.A. River bridges were built as part of an ensemble and were designed to beautify the city.(p.164)

164

99???

Neither the Asian owner nor the Hispanic customer could venture a guess concerning the origin of the name of this small Monterey Park market.

Pomona and Rosemead

Donuts, the perennial favorite.

El Monte

Automobiles were initially sold out of livery stables and blacksmith shops. As business shifted from horse-with-carriage to horseless carriage in the first decade of the twentieth century, some fledgling automobile agencies built lavish showrooms. More austere times such as the Depression and WWII, as well as the increase of working class customers, spawned simpler but still attractive auto showrooms such as the Cavalier.

Ontario

Once a "grinder" now a deli sandwich, this large neon sandwich has been luring hungry customers off the road for decades.

A Peaceful Stretch

Jets thunder by barely above tree level and the area's population has mushroomed exponentially, yet this little piece of old Highway 99 along Holt Boulevard near Ontario Airport projects a nice feeling of "the old days."

Bloomington

The Bloomington Garage was a Highway 99 landmark that stood proudly from around 1910 until 1997 when it was removed to be replaced by a new gas station/ minimart. Fortunately, some fast thinking local enthusiasts raised the thousands of dollars needed to move the historic garage and the proprietor's adjacent Sears Roebuck Kit house. They plan to relocate them to an historic park and turn it into a museum.

Colton

An oversized bag of cement stands near ever-diminishing Mt. Slover. As the cement mountain shrank, the ribbons of highway grew longer and wider.

Redlands

Drive in theaters are a roadside enterprise that first emerged in New Jersey in 1933. Their biggest year was 1958 at which time there were over 4,000 in operation. Fewer than 1,000 are left. Movie going habits have changed, and in many places the real estate is too valuable to justify the seasonal usage, even with the addition of wintertime flea markets. The Tri-City in Redlands sports a painting of a ski scene that might seem out of place in an area where orange trees grow, but snowy peaks do indeed dot the horizon on a clear winter day.

Citrus groves and wind machines at one time dominated the view along both sides of Highway 99 through this part of California. Only a fraction remain.

An example of a paper orange crate label from the citrus packing house heyday. The once disposable colorful labels are now valuable collectables.

Beaumont

The Beaumont Hotel was once a favorite destination for those seeking a dry, healthful climate, or coming out to enjoy the profusion of spring blossoms in the surrounding cherry, apple, almond, peach, and plum orchards.

Banning

How I would love to try a cup of Ritz Ice Cream for a nickel, but alas the tiny square building is the only reminder of what must have been a rich and cool treat on a summer's day.

Cabazon

A concrete bridge in a style that seems to imitate wooden posts and rails crosses a desert wash just a few feet away from the modern freeway structure that replaced it.

Blooming Yuccas

Since the start of its course through California, Highway 99 has traversed mountain and valley and metropolis. The final segment crosses part of the Mojave Desert and ends at the Mexican border.

Whitewater

The river is an unexpected treat and the 154 foot long bridge one of the nicest we've seen. The mile long loop of old highway went through the small town of Whitewater, a popular service stop under a grove of cottonwoods.

Incognito

This bumpy mile-long stretch of cement road has been innundated by an amazing array of power lines. Few people realize that this forgotten fragment was at one time part of US Highway 99.

Surreal Spectacle

Wind generating machines dot the hillsides along breezy San Gorgonio Pass. The first attempt to harness the free energy was with the Oliver Power Generator in 1926, but that effort was most likely part of a money grabbing scheme and not fully legitimate.

Date Capital

Date palms were introduced to this area in 1912 by the Department of Agriculture. Valerie Jean (named after founder Russ Nichol's daughter) has been in business since 1928. Along with Harry and David of Fruit-of-the-Month Club fame (which coincidentally, is also situated on Highway 99, in Medford, OR) this enterprise was one of the first successful mail order food businesses. When mail orders decreased during WWII, soldiers from the nearby bases helped take up the slack with their fondness for date shakes.

The End and the Beginning

Highway 99 ends and begins in Calexico just across the street from the original border crossing in front of this small town department store that has been a Calexico fixture for many years. Nary a word of English could be heard inside the busy store, crowded as it was with shoppers from across the border.

From this point, US Highway 99 went north about 1,600 miles all the way to the Canadian border at Blaine, Washington. The highway's length grew shorter over the years as various sections of road were straightened or realigned.

183

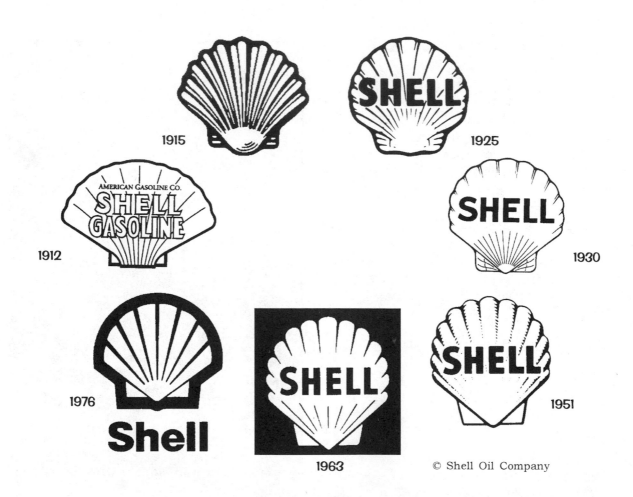

1915

1925

1912

1930

1976

Shell

1963

1951

© Shell Oil Company

184

PART

3

Appendix: Following the 99 Trail

    Our trip goes from north to south. Piecing together the fragments of the old highway is not always easy. What is labeled "99" might not actually be the vintage 99 we that we seek.

    Keep in mind that there were basically three versions of the highway: the narrow concrete road poured in the teens, the two and three-lane version of the 20s to 40s, and then the four-lane "expressway" version of the 50s that in some areas evolved into the present freeway. The earliest highway usually followed the railroad tracks, except in places where the grade was too steep for trains.

    We find the old highway remnants in a variety of forms. This can be confusing! Much of the old has been covered up by the new. When our travels put us on the freeway, it is generally the case that the old highway has been subsumed by it.

    A lot remains to be discovered in the form of "business routes" through towns that were later bypassed by expressway or freeway. Some intact segments of the old highway retain the "99" number although the road was downgraded from Federal to State Highway status in the 1960s. Conversely, some of current State Highway 99 indeed is not on the original alignment.

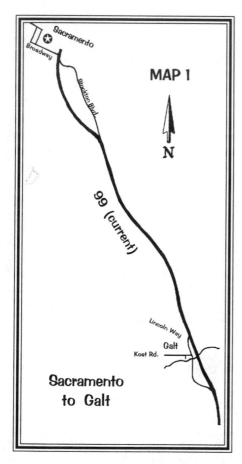

MAP 1

N

99 (current)

Sacramento

Broadway

Stockton Blvd.

Lincoln Way

Galt

Kost Rd.

**Sacramento
to Galt**

There are segments that have been given new names such, as Golden State Highway and Motel Drive. Then there are the short unnamed and forgotten segments.

Where all has been covered with asphalt and no trace of the original concrete is visible, you can still see vestiges of the past in the form of old auto courts, motels, and other roadside businesses. Where not too built up, the countryside still provides much of the same scenery that was enjoyed by travelers in days gone by. A little imagination and exploration is all it takes to travel back in time.

NOTE: Listed page numbers refer to the related photos.

Valley Route

We begin our trip down California at the Capitol Mall on the west side of the Capitol building in **Sacramento** (p.112, 113). From this juncture we can look straight west across the beautiful Tower Bridge (which carried 99W heading north toward Oregon), or east to the Capitol and Capitol Park.

Time to move on. We turn our cars south on 3rd St. (which is now a one-way street—we would be on 5th St. if we were coming from the south) and go all the way to Broadway, then turn left (east). Go under, not on the 99 freeway. This is not the 99 we are seeking.

Continue to Stockton Blvd. (p.18), which was 99 until the freeway version opened in 1962. Turn right (south). Some old highway businesses remain (motels, etc.) In about five miles there is no choice but to enter the freeway. The old highway is now covered by the freeway. The scene turns rural.

Exit at Walnut Ave. in **Galt** (p.137), turn left on W. Stockton Blvd., and go through Galt on Lincoln Way, the old highway. Galt was avoided with a short bypass early on. At the south end of Galt at the county line we encounter our first bit of virgin concrete, a nice remnant of the first state highway of the teens (p.92). Turn right onto Kost Rd. then take an immediate left. The isolated concrete segment crosses a field and leads up to a bridge site (no more bridge) over Dry Creek. Stay on Lincoln Way to Woodson Rd., then back on the freeway.

Exit at **Lodi** (p.138) on Turner Rd. Through Lodi the old highway is the "business route," Cherokee Lane. To find the Lodi Arch (p.116) turn right (west) on Pine St. and go a few blocks to the railroad tracks. Near the arch is a nice old train depot, a great loaf of bread ad painted on the side of a building, and the town's unique water tower. Back to Cherokee and onto the freeway at Century Blvd.

The southbound freeway lanes cover old 99 through here. Many old highway businesses line the frontage roads between Lodi and Stockton. The "roadside attraction" of Pollardville (p.104) is off the northbound frontage road.

To access the old highway through **Stockton** (p.27, 117), go through town on Wilson Way. The "T" at Charter Way is where 99E and 99W split for a time in the 30s, to join again in Manteca.

For 99W turn right (west) then left (south) on French Camp Turnpike. This soon joins I-5, then off the freeway at Hwy. 120, and soon a left onto Yosemite Ave. and on into Manteca. From Sacramento until this point (Yosemite Ave.) we

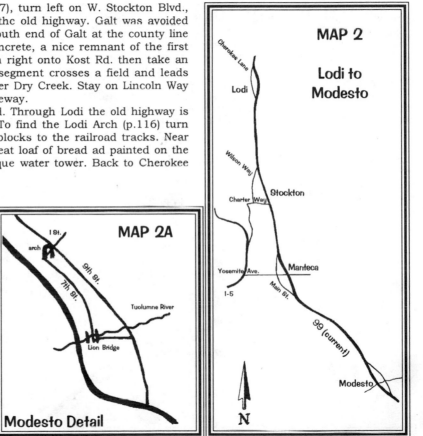

187

have been conjoined with the route of the old Lincoln Highway. Rejoin 99W at the Yosemite/Main intersection

For 99E from Stockton, left on Charter Way to Mariposa Rd., then back on the freeway.

Off at **Manteca** (p.119, 138) and through town on Main St. Where Main crosses Yosemite (where 99E and 99W converged) was a major bottleneck before the bypass was built in 1955, crowded with Bay Area tourists on their way to Yosemite National Park. Main turns into Moffat at the train tracks. The old highway follows the train tracks and the remains of the Spreckles sugar plant and large feed lot are seen on the roadside. Soon we're forced back onto the freeway.

99 remains locked under the freeway until we reach **Modesto** (p.31, 121). The old highway through Modesto is "Business 99", 9th Street. Lots of old time flavor in Modesto. Modesto, the hometown of George Lucus, was the inspiration behind *American Graffiti*. The street scenes were actually filmed in San Rafael and Petaluma. The "Water, Wealth, Contentment, Health" arch built in 1912 spans I St. at 9th (p.122).

The very earliest version of the state highway turned off 9th at I St., turned south on 7th St. and crossed the Tuolumne River on the concrete Lion Bridge built in 1916. For bridge fans, definitely worth the short detour.

Back to 9th St., we are forced on the freeway at the south end of town. About five miles later, exit at Taylor Rd. to access Golden State Blvd., Business 99. Remnant stretches of Highway 99 (the Golden State Highway) are commonly called Golden State Blvd. in the Central Valley. Golden State Blvd. takes us through and beyond **Turlock**.

Back on the freeway, through **Livingston** and on to **Atwater** (p.123). Only last year (1997) the site of the very last stoplight on Highway 99 was obliterated by the long overdue Livingston Bypass. The Foster Farms restaurant near that forced stop now stands empty.

Take Atwater Blvd. through the town of Atwater. Back to freeway south

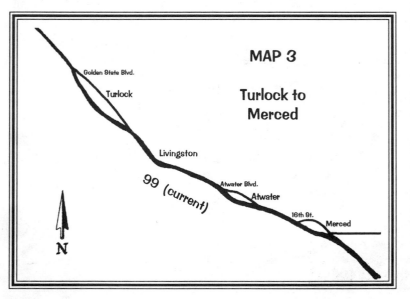

MAP 3

Turlock to Merced

Golden State Blvd.

Turlock

Livingston

99 (current)

Atwater Blvd.

Atwater

16th St.

Merced

N

of town. Next stop **Merced** (p.3, 11, 13).
Exit at Business 99, 16th St. Nice 1940
highway bridge with wooden guard rails
coming into town, Hotel Tioga, train depot.
Back on freeway south end of town.

For an interesting side trip from
Merced, go east on Highway 140 a few miles
to the so-called Agricultural Museum (so
named for permit purposes only). A
fiberglass "muffler man" stands guard
outside, the interior is filled with a
wonderful array of old toys, machines, and
household objects, the back lot has antique
gas engines and a 1915 gas station.

Next piece of the 99 alignment is
encountered at **Chowchilla**. Exit at Ave.
15, which turns into Chowchilla Blvd., the
old highway through town. Chowchilla
Blvd. discontinues at the center of town
near the central freeway exit; gas stations
and fast food restaurants now fill the space.

The old road starts up again just south
of this. It is lined with black walnut trees
and passes by a fig and an olive orchard.
Return to the freeway by turning left at the
stop sign across from the olives, for the old
alignment ends abruptly soon after this
point.

The next sight (p. 126) is the grotesque
remains of a "Mammoth Orange" cafe
(despoiled by blue paint and graffiti) at Road
19 on the west side of Highway 99 (which
through here is a four-lane divided
expressway). The Mammoth Orange farther

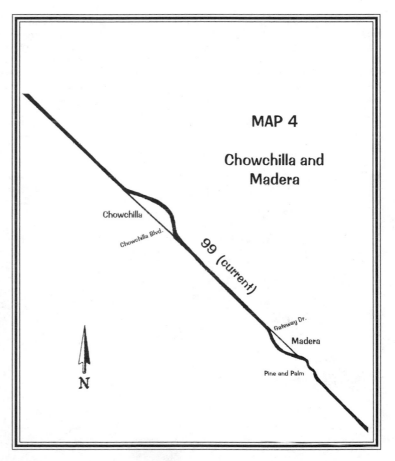

MAP 4

Chowchilla and
Madera

Chowchilla

Chowchilla Blvd.

99 (current)

Gateway Dr.

Madera

Pine and Palm

N

down on the east side at Road 22 1/2 is still thriving (p.127).

Exit at Gateway Dr., the old highway through **Madera** (p.68, 70, 128, 129). Many old motels and cabins, the Ramirez Gas Station (on the right) that was built in the mid-40s, and a nice little 1935 bridge just past it. Back to the freeway at the south end of Madera.

About a mile south of Madera in the median between the north and southbound freeway lanes look sharp to see the **pine and the palm** (p.130) which were

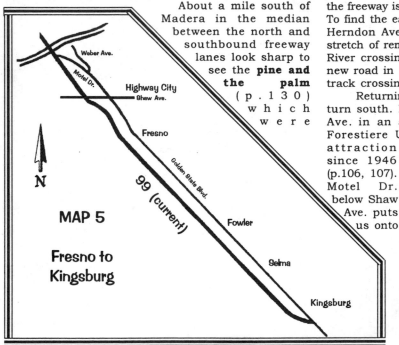

MAP 5

N

Fresno to Kingsburg

planted here near the geographic center of California to represent the meeting of north and south. For more relaxed viewing exit the freeway to the east side and drive through the vineyard (ask permission) which edges the freeway 99. That will put you only a few feet across from the trees.

The next place the old 99 alignment splits from the freeway is on Motel Dr. a few miles north of **Fresno**. To find the earliest road turn east from Motel Dr. onto Herndon Ave., then north on Weber. This mile long stretch of remnant highway led up to the San Joaquin River crossing. It was replaced by a new bridge and new road in 1928, eliminating two dangerous railroad track crossings in less than two miles.

Returning to Motel Dr. (the post 1928 highway), turn south. Half a block west of Motel Dr. on Shaw Ave. in an area known as Highway City are the Forestiere Underground Gardens, a roadside attraction since 1946 (p.106, 107). Motel Dr. below Shaw Ave. puts us onto

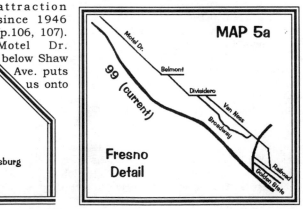

MAP 5a

Fresno Detail

the freeway. Motel Dr. reemerges at the Clinton Ave. exit. A number of 99-vintage motels line the frontage road (Parkway).

Approaching **Fresno** proper (p.69, 131, 132, 133, 134) on Motel Dr., we pass a park, then the old highway route jogs to the left at Belmont, going under the train tracks through an attractive subway. A quick right on H St., a left to Divisidero, a right on Broadway and all the way through Fresno. (Broadway is one way so if going south-to-north you will be on Van Ness, the pre-1932 highway route.) This town has a number of historical buildings including some ornate theaters and the Fresno Water Tower (p.139).

Stay on Broadway until it is no longer one-way, cut to the left to Van Ness, then right. Shortly before reaching Railroad Ave. we go right under the 1925 Van Ness arch (p.36) advertising "The Best Little City in the U.S.A." which spanned the highway route through Fresno for only a few years.

Railroad Ave., Golden State Blvd., and freeway 99 parallel each other a short distance apart and are three successive versions of Highway 99 on the south side of Fresno. In 1932 the route was changed from Van Ness onto newly improved Broadway. Broadway connected to Railroad (and later Golden State) via Cherry Ave. Going south, Railroad soon merges onto Golden State Blvd., which is intact for about twenty miles, all the way through **Fowler** (p.135), **Selma**, and on to **Kingsburg** (p.139).

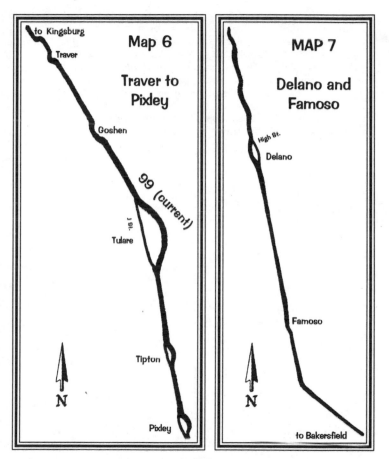

191

At Kingsburg old 99 is called Simpson St. It is cut off abruptly by the freeway when you reach the Del Monte packing plant. Backtrack and enter the freeway.

The next bit of old 99 appears at the **Tulare** exit (p.141, 143, 144). The old highway through town is J St. if heading south, K St. if heading north. Back on the freeway, and a couple other very short segments of old highway (a quick on and off the freeway) are encountered in the small towns of **Tipton**, **Pixley** (p.21), and **Earlimart**. Old 99 through **Delano** (p.145) is carried on High St.

Nearing **Bakersfield,** an isolated length of old 99, the 1933 Golden State Highway, fronts the tracks and is accessed from the Seventh Standard Rd.

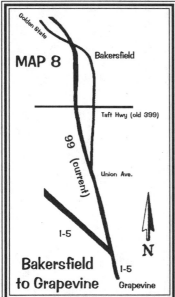

MAP 8

Golden State

Bakersfield

Taft Hwy (old 399)

99 (current)

Union Ave.

I-5

I-5

Bakersfield to Grapevine

Grapevine

N

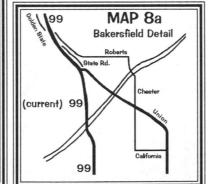

MAP 8a
Bakersfield Detail

Golden State

99

Roberts

State Rd.

Chester

(current) 99

Union

California

99

exit. Return to freeway, exit at Golden State Ave. and head a short distance north on Airport Rd. State Rd. (parallel to freeway on north side) is another fragment from the same alignment that has the old "feel" to it. This section was abandoned in the mid-50s when the Airport Rd. intersection was put in.

Back to Golden State, cross the Kern River on a nice 1933 bridge (widened from three to four lanes in 1954) with metal pipe railing. The road angles south, turns into Union Ave. and continues as such for many miles. This was the 99 route until the freeway 99 was built in phases 1962-1965. Before 1933, the 99 alignment went (N to S) on Roberts Ln. (access from Airport Rd.) to Chester, south on Chester to California, east on California to Union, and on south.

Bakersfield is full of 99-era motels and coffee shops, including the once-grand Bakersfield Inn (p.146) with its covered walkway/arch spanning Union Ave. that connected portions of the motel on both sides of the road.

The old highway in the guise of Union Ave. becomes a tree-lined boulevard with oleanders in the median south of town, a nice example of 1950s four-lane expressway, then rejoins the freeway 99 in a few miles. The Union Ave./Taft Hwy. junction in Greenfield is the old 99/399 junction. A few miles further, freeway 99 merges onto I-5, and that is the last we see of the number "99." From here on out, no portion of the former Highway 99 retains its number.

Leaving **Bakersfield**, it is a straight shot south to the foot of the **Grapevine Grade**.

Ridge Route and Ridge Route Alternate

From where Union Ave. (old 99) joins I-5 south of **Bakersfield**, the beginning of the modern rendition of the Ridge Route as it climbs up **Grapevine Grade** is quite visible. What we see today are the two widely separated directions of Interstate, southbound going up the west side of the canyon, northbound coming down the east side.

Keep in mind that between here and Gorman (the Grapevine portion of the Ridge Route) the highway from the teens and the highway from the thirties followed generally the same alignment. The result is bits and pieces of each road being found in close proximity to each other, or on top of each other.

In contrast, from Gorman south to Castaic, the two alignments were miles apart and must be explored separately.

As we look south, the 1915 Ridge Route came down the same side of Grapevine Canyon that now holds northbound I-5. Most of the 1915 road was covered up by the freeway although a few of the twists and turns can still be found, isolated by I-5 as it cuts across in a much straighter swath. The 1935 version of Highway 99 came down the west side of Grapevine Canyon and is now mostly covered by southbound I-5 (p.49, 52, 64).

A nice little remnant from the 1915 road is found even before starting up the grade. Exit the freeway at **Grapevine**, go to the east side and turn onto Grapevine Rd. We are now between the north and southbound freeway lanes and at the south end of the 17 Mile Tangent (see text) that came from Bakersfield. At .8 mile is the shell of an old auto court (p.147). Turn left, and here is where the old road started its

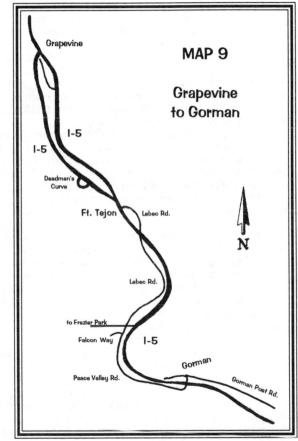

MAP 9

Grapevine to Gorman

193

twist up "the Grapevine." You can drive another .3 mile and walk a little farther up the old road (note original concrete curbing) to a vantage point directly above the freeway.

Back to I-5 and the next place to catch a glimpse of the 1915 road is just before the "Ft. Tejon Exit 1 Mile" sign. Look to the right (west) to see remnants of old concrete highway in amongst the vegetation. This was part of the infamous Deadman's Curve (p.151). We'll get a better view from our next stop.

Exit at **Ft. Tejon**, head for the fort but instead turn right onto Dieger Rd. You will be going back to the north paralleling I-5 on the dirt road. At around .9 mile look down and Deadman's Curve is highly visible between Dieger Rd. and the freeway. Consider the amount of dirt removal necessary to make the huge freeway cut that avoids this sharp curve. We've come a long way!

Return to the freeway but do not enter. Cross over to the east side to Lebec Rd., parts of which are the 1915 alignment. Bits of concrete can be seen peeking from under the asphalt between the I-5 overcrossing and **Lebec** School. At the school, notice the school buses parked on concrete; this too was part of the old highway. Shortly thereafter, another concrete section appears in a field (behind a private gate) off to the right and angles into (is cut off by) the freeway.

Still on Lebec Rd., we cross over the freeway again and just past the Mobil Oil plant the 1915 alignment reappears. It has been covered with asphalt but some concrete patches can be seen, as well as original concrete curbs. Continue on Lebec Rd. to the Frazier Park exit, where Flying J Truck Stop is located. Up to this point, all we've seen is 1915 Ridge Route remnants as the Ridge Route Alternate of the 1930s has been covered by I-5 from here down to Grapevine.

At the **Frazier Park** intersection, Lebec Rd. turns into Peace Valley Rd. Continue on. Now we are on the 1933-35 Alternate, which is one half of the 1951 four-lane expressway.

For a view of Ridge Route evolution from one setting take a right on Falcon Way a short distance beyond the Frazier Park intersection. The gated road leads to a high school and is sometimes locked. Just up Falcon Way a concrete piece of the old highway comes in on the left. A short walk up this takes us to an abrupt end (concrete slab sheared off) at the top of a shallow cut above Peace Valley Rd. (If the gate at Falcon Way is locked, it is a short walk up the bank from Peace Valley Rd. at a point .1 mile south of **Falcon Way/Peace Valley** intersection.)

From the vantage point of this 1923 double slab (which was laid directly on top of the original 1919 slab), the 1951 four lane expressway (Peace Valley Rd.) and I-5 are visible. The 1930s Ridge Route Alternate was completely obliterated, being too windy to be incorporated into the 1951 alignment.

Continuing south on **Peace Valley Rd.**, we are following both the 1915 and 1933-35 alignments. Topping Tejon Summit, you will notice that I-5 just to the east crests the summit at a lower elevation; incredible amounts of material were moved during construction to decrease the slope, and the elevation of the summit was considerably lowered.

Still on Peace Valley Rd., we head down toward **Gorman** still on the '33-'35 (and subsumed 1915) alignment. After crossing the turnoff to Hungry Valley

Recreation Area we can see where the old highway angles off of Peace Valley Rd. back across the freeway to emerge on the other side. Stuck on Peace Valley Road, we loop around a little to the west and into **Gorman**. To access that piece across the freeway, go into Gorman, under the freeway to Gorman School Rd./Gorman Post Rd. intersection. Turn back north. The 3-lane stretch of concrete goes for .5 mile and is then cut off by I-5.

Turn around and go back south on Gorman Post Rd. At this point we are following the 1915 and the 1933 alignments of the Ridge Route (Highway 99) as well as an old section of Highway 138. (138 goes east to the desert and was where one version of the pre-Ridge Route "Midway Route" that bypassed these mountains joined the old wagon road to Ft. Tejon). In about five miles we must turn onto a more modern Highway 138, but don't make the turn just yet. Right in front of us is another nice but short section of concrete highway that is quickly cut off by the current 138.

Near this point is where the two early versions of 99 diverged. The 1915 Ridge Route went off to the east, and the 1933 Alternate (p.60, 63) went off to the west, to join again in Castaic. By making a loop trip, both routes can be covered. We will start with the Alternate, of which much less remains. Most of it was either incorporated into I-5 or sunk beneath Pyramid Lake.

Turn right on 138 and return to freeway. We find the next piece of exposed Ridge Route Alternate, which is also the postwar expressway, at Smokey Bear Road on the west side of the freeway. Turning right (north), it goes for 1/4 mile then ends at freeway's edge. Turning left, it goes for 1.5 mile to the Pyramid Lake entrance. They might let you drive in without paying and have a

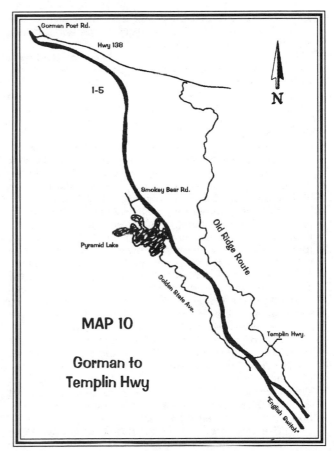

MAP 10

Gorman to
Templin Hwy

look—the old roadbed goes right into the lake, and one of the best parts of Highway 99 is hidden for the next ten miles. We are at the north end of Piru Gorge.

Back to I-5 and off again at **Templin Highway**. On the west side we find a fairly lengthy stretch of the four-lane expressway (called Golden State Highway) parallel to the freeway, one sure to summon memories of 1950s Ridge Route crossings. Turning left, the road is quickly cut off, but turning right we travel for several miles (p.155).

Here the road is divided for a short distance and soon we reach Whitaker Summit at the top of the Five Mile Grade (to the south) and Three Mile Grade (to the north.) At Frenchman's Flat, a favorite fishing and relaxation spot then and now, we reach Piru Creek as it turns toward the ocean, and here the old highway is blocked off (cover photo). It is a pleasant walk of about two miles beside the creek to the bottom of the dam (built in 1970) and the end of the road. We pass a nice rock roadside rest area (where radiators were filled from a fountain), cross a bridge. Note the 1933 half and the 1951 (expressway-era) addition. At the end of the road is the dam and Pyramid Rock, formed accidentally when the road was blasted through here in 1933. A scene from the movie *Breakdown* was filmed here.

Returning to I-5, most of the rest of the Alternate is covered by freeway. Going down the Five Mile Grade to **Castaic**, the northbound lanes follow the old alignment. When I-5 was built, the two directions were separated and switched sides, northbound to the west and southbound to the east (the "English Switch") to take better advantage of the grade.

The last piece of Ridge Route Alternate is in **Castaic**. Exit the freeway, go to Castaic Rd. and turn left. A piece of three lane concrete highway goes for about a mile north and ends at the freeway. South on Castaic Road to Ridge Route/Castaic Road intersection, and here is where the Ridge Route and the Ridge Route Alternate split on the south end. The old highway continues a short distance south and again is cut off by I-5.

Return to the Ridge Route/Castaic Rd. intersection for the trip across the original 1915 Ridge Route.

1915 Ridge Route

The Ridge Route, so quickly forgotten when the Alternate opened in 1933, is easy to follow (it being more or less intact) but is slow going. Most of the way, speeds not much faster then the original 15 mph speed limit can be achieved (p.8, 59, 154).

In this one case we travel south to north, in keeping with the loop trip concept. We start at the Ridge Route/Castaic Road junction (Map 11). In about seven miles (Map 10) we reach a stop sign at Templin Highway (built in 1968 to access a power plant) which cuts deeply across the old road bed. Crossing that, we enter the portion of the Ridge Route which in 1997 was entered into the National Register of Historic Places. The integrity of the road has been compromised in a few places (mostly by laying gas pipelines) but one surely gets the feel of highway travel early in the century by driving it. The concrete twists and turns and the 1920s "improvements" (curves widened and filled in with asphalt) are easily discernible. Remnants of roadside businesses are still visible.

Mileages are from the Templin Highway crossing (and probably are not exact). Five miles on the left sat the white clapboard National Forest Inn. Not much remains but the cement steps and foundations and some obviously planted trees (locust and cypress.) A fire recently burned through here, not an uncommon occurrence in this country. Not far past this there is a very good view way down and off to the west of I-5 and the Three Mile Grade portion of the Alternate.

At eight miles, we reach Swede's Cut (p.56), the first portion of the Culebra Excavations where power shovels were used (a rarity). We go through more road cuts and to Reservoir Summit at mile ten. A spur road on the left goes up to the site of the restaurant, store, and gas station. There are some planted pines (we are still in brush country), cement foundations and the cement reservoir that was built during the highway construction.

At 12.5 miles was Kelly's Half Way Inn on the right. Not much but a flat area and a single cypress tree. At 14.8 miles was the Tumble Inn (p.153), another hotel and garage. The native rock foundation is about the nicest man-made artifact along the route. "Tumble Inn" is etched into the concrete.

At 16.7 is a large rock outcrop known as the Granite Gate (p.152), and at 17.2 was Horseshoe Bend rounding steep Liebre Gulch. These were the types of landmarks that the early motorists looked for and were depicted in 1920s postcards.

The brush begins to recede, oaks and pines appear. The site of the three story log and stone Sandburg's Summit Hotel (at 18.6 miles on the left) has cement foundations, rock walls, and is shaded by stupendous oaks. A little further on is a wonderful desert view of Antelope Valley. This truly is a "ridge route."

At 21.3 miles we reach Highway 138 (which covers the remainder of the Ridge Route between here and Gorman Post Road.) Return west to the freeway.

Santa Clarita Valley

The journey south resumes back at the Castaic/ Ridge Route intersection on the east side of I-5. Go back over the freeway to **The Old Road**. Parts of this are indeed "the old road" but not all of it and not right here. Going south toward **Castaic Junction** and L.A. on The Old Road, old 99 soon emerges again from under the freeway. We cross Castaic Creek on a 1932 bridge. Remnants of the earlier bridge (p.157) can be seen below on the right (west).

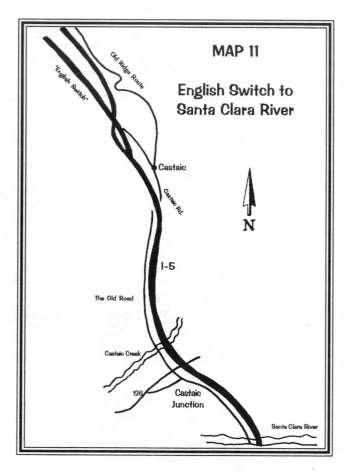

MAP 11

English Switch to
Santa Clara River

N

Old Ridge Route

"English Switch"

Castaic

Castaic Rd.

I-5

The Old Road

Castaic Creek

126

Castaic
Junction

Santa Clara River

Castaic Junction is where 99 and Highway 126 from the coast crossed. This was once a major stopover as evidenced by the large shady parking lot that used to fill with trucks and cars patronizing the gas stations and restaurants before (or after) the mountain crossing. A pristine tree-lined section of the old 126 (the current 126 being an adjacent freeway) goes a short distance west.

The Old Road (99) continues south from Castaic Junction as a divided road bisected with oleanders. We cross the Santa Clara River on an interesting bridge (dated 1928, rebuilt after the flood - see text p.74) with latticework sides.

At Magic Mountain Parkway (formerly Saugus Rd.) we again come to a place where Highway 99 had a major realignment. Before the Weldon Canyon Bypass was completed in 1930, 99 went through **Saugus** and **Newhall**. To take that route, follow Magic Mountain Parkway to San Fernando Rd. and turn right (south.) Continue to the Sierra Highway, right again and rejoin The Old Road.

The Sierra Highway portion of Highway 99 also carried old Highway 6. This was the south end of the Midway Route that, before the Ridge Route was built, went around the mountains, through the desert and then back to Gorman and north to Bakersfield. On this very old road, initially built for wagons, is Beale's Cut (p.158) and the Newhall Tunnel (p.76) site. These are easily accessed from the post-1930 alignment as well, which we will later do.

Resuming our position at the junction of **The Old Road/Magic Mountain Parkway** and driving south, the post-1930, pre-freeway Highway 99 is soon covered by I-5. Enter the freeway at **Valencia Blvd.**, then exit at **Calgrove**. 99 appears again on the west side (still as The Old Road) then crosses over to the east side. These I-5 freeway bridges collapsed in a 1994 earthquake and this piece of old Highway 99 was invaluable as a detour. We top Weldon Summit, go down and soon arrive at the Old Road/Sierra Highway junction (p.159).

The beautiful curved concrete bridge (p.78) at this juncture carried Highway 99, in its Newhall and Saugus routing, until the 1930 realignment. To find Beale's Cut (worth the trip and short hike) go up the Sierra Highway .8 mile to the historical markers on the right, then walk east (5 or 10 minutes) on a well defined trail. Newhall Tunnel was located at the now very wide Newhall Pass another mile out the Sierra Highway.

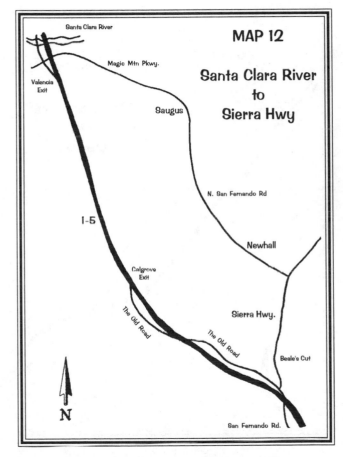

MAP 12

Santa Clara River to Sierra Hwy

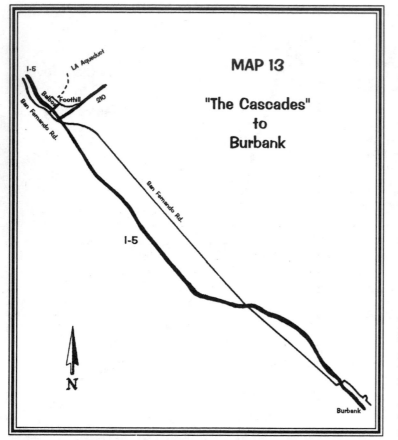

MAP 13

"The Cascades"
to
Burbank

I-5

LA Aqueduct

Balboa

Foothill

210

San Fernando Rd.

San Fernando Rd.

I-5

Burbank

N

Back to the intersection, The Old Road now is called San Fernando Rd. (not the same one that went through Newhall) and continues to L.A. The original Golden State Freeway (signed as 99 at the time) was pushed through here in 1954, leaving the earlier alignment intact. Segments of this early freeway are incorporated into the I-5 truck lanes at the 5/14/210 interchange. The original 3 level interchange is still at the Sierra Highway overcrossing.

Back to the old highway (San Fernando Rd.), we traverse San Fernando Valley. The North Burbank Overheads were built in 1941 and are the only major railroad crossing in the San Fernando Valley. Train tracks and industry. Motels, car washes, cafes. The towns of **San Fernando, Pacoima, Burbank, Glendale** (p.161, 162).

Los Angeles

The course of Highway 99 through **Los Angeles** changed several times as traffic patterns evolved and roads were improved. This makes it difficult to trace. I used maps from 1926, '27, '30, '34, '39, '41, and 1950 to reconstruct the routes, but this log may not be 100% accurate.

It is definite that the earliest routes went through the heart of the downtown. To follow the earliest labeled route, stay on San Fernando past where it turns into

Avenue 20 and go to Broadway. Turn right and cross the L.A. River on the 1910 North Broadway (formerly the Buena Vista) Bridge, unfortunately shorn of its original ornamentation. Broadway goes through Chinatown (p.82), City Hall in the distance. Left on Ceasar Chavez (called Macy St. in the 99 days) past El Pueblo de Los Angeles (Olvera St). In three blocks (with Union Station [p.163] just across the street) left again on North Main and back across the river on another 1910 bridge. The Los Angeles River bridges invite further exploration (p.164).

New research indicates there was another accepted early highway route into downtown that left San Fernando Rd. north of Burbank at Lankershim Blvd., to Hollywood Blvd., to Western Ave., to Wilshire, to Broadway. Even the "experts" have trouble untangling this web.

Leaving Los Angeles, the north/south Highway 99 goes distinctly east-west for the next fifty or so miles.

MAP 14

Through Los Angeles

N

San Fernando Rd.

I-5

110

101

Daly St.

Broadway

Main

Main

Valley Blvd.

I-10

Valley Blvd.

Fremont St.

Garvey Ave.

After going under I-5 and shortly thereafter crossing Mission, Main St. turns into Valley Blvd. Valley carried Highway 99 south (actually east) until 1935.

By 1936 a new course was taken. Highway 99 left San Fernando Rd. and shared billing with Route 66 going down Figueroa St., which intersected north of Broadway, crossing the river on the 1936 Figueroa St. Bridge. The bridge is still there but Figueroa St. west of the river, the portion that carried 99, is now covered by the Pasadena Freeway (110).

99 then turned east at Sunset, splitting from 66 and joining 101 for a few blocks. Sunset turned into Macy (Ceasar Chavez) St. at Broadway, where the earlier 99 came in. Then, right on Alameda, passing right in front of Union Station, and left on Aliso St., now covered by the Santa Ana Freeway. 99 crossed the river and headed out on Ramona Blvd.

It is likely that part of the reason for the 1935 rerouting (away from Broadway and Main) was the reconstruction of Ramona Blvd., the extension of Aliso on the east side of the river. In 1935 Ramona Blvd. was made into a six-mile "airline urban route" (as one magazine called it) designed to speed traffic into and out of the city from the east. A few years later the Aliso Street Bridge over the river was torn down and rebuilt in order to accommodate ever more traffic and the planned freeways. In 1943 Ramona Blvd. became Ramona Parkway (later the San Bernardino Freeway) and still carried 99.

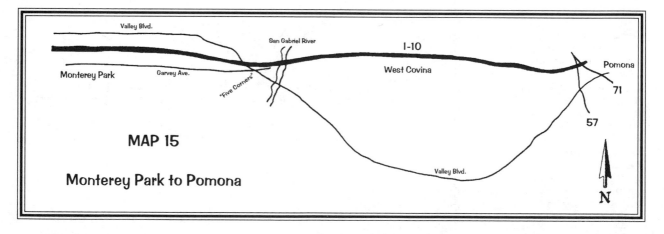

MAP 15

Monterey Park to Pomona

When that freeway was later extended west from the river, Highway 99 was routed through the famous four-level interchange west of downtown for its jog east and north. (That is, 99 coming from the north, as part of the Pasadena Freeway, switched on to the San Bernardino Freeway via this structure.) When the Golden State Freeway (I-5) was opened coming into L.A. from the north in 1962, Highway 99 was moved to that alignment and no longer went through downtown or the four-level interchange. By 1964, the "99" signs were removed from that portion of I-5.

There is some evidence of an alternate 99 route (pre-Golden State Freeway) that avoided the downtown area as well. That left San Fernando Rd. to go down Ave. 26, Daly St. and Marengo St. to connect with Ramona.

The Inland Empire

During the period when Highway 99 left L.A. by crossing the Main St. Bridge (previous to 1935), 99 went on Valley Blvd. all the way to **Pomona**. That rather indirect route is still intact, following the train tracks through industrial areas. Main St. turns into Valley Blvd. at Mission, and Valley goes all the way to Pomona and connects to Holt Blvd. This is where the pre- and post-'35 routes between L.A. and Pomona met.

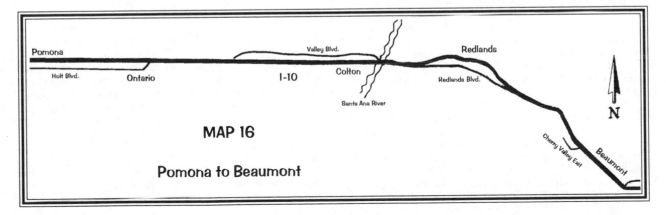

MAP 16

Pomona to Beaumont

After 1935, Highway 99 left L.A. via Ramona Blvd. and then made a right angle turn onto Garvey Ave. That sharp turn is intact but between there and L.A. all of Ramona has been covered by freeway. To find this spot, if on Valley, cut down on Fremont to Garvey. Or take the Atlantic exit off of I-10 and go south to Garvey. In either case, backtrack west to find the sharp turn, after which the old road is shortly cut off.

Going back east, Garvey is intact for several miles, up to the San Gabriel River. Just before that point, Valley crosses Garvey at a big intersection that was commonly known as Five Corners. The road passes through what were once pleasant, outlying little towns. **Monterey Park, Rosemead, El Monte, West Covina** (p.165, 166, 167).Lots of motels and trailer parks with good mountain views (when there's no smog.)

East of the river, a new version of Garvey continues as a freeway frontage road but all of the Garvey that was 99 is covered. 99 reappears as Holt Blvd. in Pomona.

Holt is easily followed through **Pomona** and **Ontario** (p.91, 168, 169). Just past Ontario Airport is a pretty, eucalyptus-lined divided piece of old highway, then Holt is covered by I-10. Picturesque **Guasti**, which used to boast "the largest vineyard in the world," is a short side trip.

Enter the freeway going east then exit a couple of miles later at Valley Blvd. (Yes, another Valley Blvd.) This 99 remnant goes through **Fontana, Bloomington** and **Colton** (p.171) then ends. Return to the freeway to cross the Santa Ana River. The westbound I-10 freeway bridge is actually the old 99 alignment.

After crossing the river, exit at Waterman South and go to Redlands Blvd., turn left. This piece of 99 passes through **Redlands**, a pleasant town (p.85, 172, 173). There are even some citrus groves still in the area. We must return to the freeway on the far side of Redlands.

To the Border

At the Cherry Valley exit, a surprise: the first remnant of concrete highway we've seen since Castaic. After exiting, turn right (southwest), than a quick right on the frontage road. This goes for .6 mile and ends at a junk store with old cars parked on the old concrete highway just before it angles into the freeway. Backtrack to I-10.

We next encounter old 99 at **Beaumont** (p.174). Take the first exit and go through town on 6th St. At the other side of town, 6th becomes Ramsey. Follow Ramsey all the way to **Banning** (p.175). Both of these towns have a small town, old time feel to them. Back to the freeway.

Exit next at Fields Rd., turn right and parallel the freeway going back the way we came. This piece of old 99 wedged between the freeway and the railroad tracks goes for 1.8 miles and crosses two bridges, one of them quite pleasing (p.176). Turn around but do not reenter the freeway. Stay on this frontage road (still the old highway) all the way into **Cabazon**. It is a wide boulevard on approaching town, now called Main St. The famous dinosaurs (p.108) not of the 99 era but still fun are found across the freeway at the Wheel Inn.

Returning to the freeway, we go through the **San Gorgonio Pass**, full of wind generators (p.180), and begin the descent to the desert. At the **Whitewater** exit we encounter one of the nicest pieces of old 99 in southern California. The old highway bends away from the freeway for about a mile to cross the Whitewater River on an eight-spanned, 154 foot long concrete bridge built in 1923 (p.178). The beauty of the setting is a pleasant surprise. Backtrack to freeway.

Our next exit is Indian Ave. Turn south, then left on **Garnet Ave.**, another short concrete section. Not particularly beautiful, festooned with an incredible number of power lines (p.179), it is there nevertheless. Backtrack to freeway.

Next exit, Ramon Rd. Go over to **Varner Rd.** on north side of freeway, which follows the old alignment all the way to Indio. Bits of cement are seen peeking through the asphalt.

At **Indio** (p.71), the Jefferson St. offramp connects with **Indio Blvd.**, the old highway. Cross a nice double bridge (4-lane era) over the Coachella storm drain, and through Indio we go. Clark's Truck Stop dates from the 99 era (1947 to be exact.) A beautiful mural on the side of the building celebrates the early days of Indio (and Highway 99) and inside there is even a Highway 99 Museum.

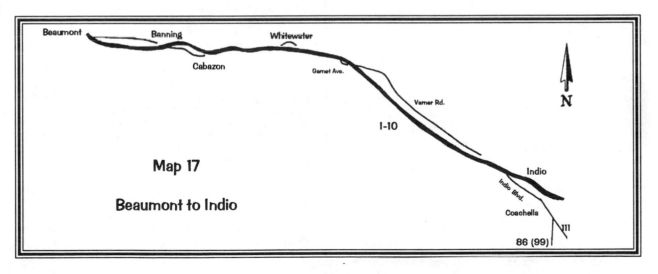

205

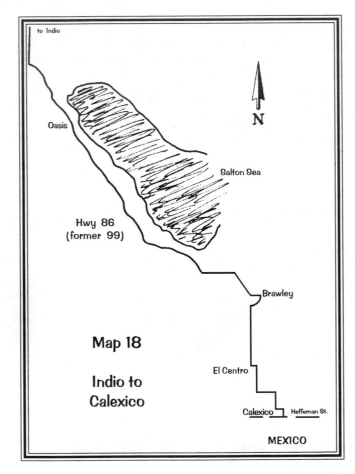

to Indio

Oasis

Salton Sea

Hwy 86
(former 99)

Brawley

Map 18

**Indio to
Calexico**

El Centro

Calexico ⌐ Heffernan St.

MEXICO

N

On the far side of Indio, Indio Blvd. becomes **Highway 86**. 86 follows the old 99 alignment to within five miles of the border. Some two-lane sections with a 99 "feel" remain; other sections are four-lane.

After Indio and **Coachella**, old 99 finally turns sharply south and skirts the western shore of the **Salton Sea** for nearly 40 miles, passing through an area of date palm groves (p.99, 181, 182) . We are below sea level for the rest of our journey. **Oasis** was a favorite stop for a cold drink or a date shake.

Then, the **Imperial Valley**, fields of crops being harvested even in the dead of winter, 125 degree weather in the summer. **Brawley**, **El Centro** and **Heber** maintain a small town flavor. In El Centro, 99 joined with Highway 80 for a few blocks (from where 86 turns onto Adams to where it crosses Main.) That east-west highway connected San Diego to Yuma, Arizona and was replaced by I-8.

86 ends a couple of miles past Heber. Turn right (south) on **111** for the final five mile trip to the end. Cross the All-American Canal and we're in **Calexico**. 111 heads for the border crossing, but the earlier crossing was two blocks to the east. To find it, turn left on 3rd, then right on **Heffernan**, and to the chain link fence. The Old Customs House is on the National Register of Historic Places. The Sam Ellis Store (p.183) is a small town department store frozen in 1950, but all the patrons speak Spanish. An "Historic Route 99/Begin" sign is just outside the door.

Note: Road segments have been found and identified to the best of our abilities.100% accuracy is unlikely. Please send comments and corrections to Living Gold Press.

Bruce Clark from Clark's Truck Stop in Indio is proud of the beautiful mural on the side of his building. The painting highlights local history, in which Highway 99 played a prominent part. Inside the store is a Highway 99 museum.

BIBLIOGRAPHY

Baeder, John. *Gas, Food, and Lodging.* Abbeville Press, New York, 1982.

Belasco, Warren James. *Americans on the Road: From Autocamp to Motel, 1910-1945.* MIT Press, Cambridge, MA, 1979

Buckley, Patricia R. *Highway 99: A California Chronicle.* Self-published, 1987.

Buckley, Patricia R. *Those Unforgettable Giant Oranges.* Self-published, 1987

Boudier, William. *The Paths of Humanity.* CA Division of Highways, 1966.

Built in the U.S.A. National Trust for Historic Preservation, Preservation Press, Washington D.C., 1985

California Highways and Public Works. California Department of Transportation, various issues.

Dasmann, Raymond F. *The Destruction of California.* Collier Books, NY, NY, 1966.

Donley, Michael, et al. *Atlas of California.* Pacific Book Center, Culver City, CA, 1979.

Drury, Aubrey. *California, An Intimate Guide.* Harper and Bros., 1939.

Fireman's Fund of California. *Automobile Tour Book of California, 1918.*

Hamilton, E. E. *Thorpe's Illustrated Road Map and Tour Book of California,* 1911.

Hanson, Harry, editor. *California, A Guide to the Golden State.* American Guide Series. Federal Writer's Project, Hastings House, NY, 1939.

Historic Highway Bridges of California. California Department of Transportation, 1990.

International Pacific Highways System.. Automobile Club of Southern California, Los Angeles, CA,1934.

Leadabrand, Russ. "Around the Old Ridge Route." *Westways* March 1963:20-23.

Longest Stretch of Improved Road." US Dept. Of Agriculture Press Release, December 16, 1928.

Milburn, Douglas H. and Harrison I. Scott. "National Register of Historic Places Registration Form for Old Ridge Route," filed 5/1/96.

Osborne, Willis. "The Old Ridge Route." *Dogtown Territorial Quarterly*, [month unknown}, 1996: 12-19.

Pittman, Ruth. *Roadside History of California*. Mountain Press Publishing Company, Missoula, Montana, 1995.

Protteau, Lyn. "The Ridge Route." *Action Era Vehicle*, July-August, 1984: 6-8.

Robinson, John W. "The Old Ridge Route, Los Angeles to Bakersfield the Hard Way." *The Branding Iron* Spring 1986: 1-9.

Rolle, Andrew F. *California, A History*. Thomas Y. Crowell Company, New York, 1963.

Root, Norman. "A Century of Good Roads," unpublished paper, CalTrans Centennial Coordinating Committee, Nov. 17, 1994.

"The Romance of the Ridge." *National Motorist* November, 1933: 9-10.

Schoenherr, Allan. *A Natural History of California*. University of California Press, Berkeley, 1992.

Suter, Coral. "Riding High on the Old Ridge Route." *The Californias* March/April 1993:18-31.

Weingroff, Richard F. "Federal-Aid Highway Act of 1956: Creating the Interstate System." *Public Roads On-Line*, Summer 1996.

Winn, Bernard. *Arch Rivals: 90 Years of Welcome Arches in Small-Town America*. Incline Press, Enumclaw, WA,1993.
Web sites:
    "Historic Auto Tour of US 99"   www.smartlink.net/~mapmaker/us99000.htm
    "Clark's Auto/Truck Stop"   www.websites2000.net/clarks/
    "Historic US 99"   www.gbcnet.com/ushighways/US99.html
Correspondence with:
    John Fisher, City of Los Angeles Dept. of Transportation
    Willis A. Osborne, retired teacher of 44 years
    Mike Ballard, youthful old highway expert
    Readers Scott Irby (San Jose), James Martin and John Sweetser (Fresno)
    Vivian Davies, Patrick Frank, and Doug Pruitt, Route 99 Association of California

To Order Books

- **That Ribbon of Highway:** *Highway 99 from the Oregon Border to the State Capital*
  $14.99, 176 pages

- **That Ribbon of Highway II:** *Highway 99 from the State Capital to the Mexican Border*
  $15.99, 216 pages

- Recollections from my Time in the Indian Service 1935-1943 ♦ Maria Martinez Makes Pottery by Alfreda Ward Maloof
  $15.00, 64 pages 8"x9"

- Shipping: Priority Mail: $3.75 one or two books
  Book Rate: $2.50 one or two books
  California residents add 7.25% sales tax

LIVING GOLD PRESS

P.O. Box 2
Klamath River
CA 96050
fax: 530-465-2371
www.livinggoldpress.com
jandk@livinggoldpress.com

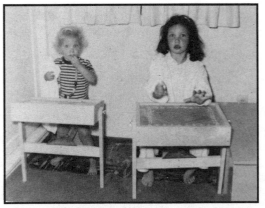

Authors Jill and Kathryn,
Living Gold Press Headquarters, 1954.

210